"ALL WE DID WAS FLY TO THE MOON"

"High Flight"

John Gillespie Magee, Jr.

Oh, I have slipped the surly bonds of earth
And danced the skies on laughter-silvered wings;
Sunward I've climbed, and joined the tumbling mirth
Of sun-split clouds—and done a hundred things
You have not dreamed of—wheeled and soared and
swung
High in the sunlit silence. Hov'ring there,
I've chased the shouting wind along, and flung
My eager craft through footless halls of air.
Up, up the long, delirious, burning blue
I've topped the windswept heights with easy grace
Where never lark, or even eagle flew
And, while with silent, lifting mind I've trod
The high untrespassed sanctity of space,
Put out my hand, and touched the face of God.

"John Gillespie Magee, Jr. was an American who joined the Royal Canadian Air Force and was killed at the age of nineteen, flying a Spitfire during the Battle of Britain. He was born in China and lived in Washington, where his father was rector of St. John's Church across from the White House. "High Flight" is without a doubt the best-known poem among aviators and can be found, elegantly framed or carelessly tacked, on home and office walls around most military airfields." —*Michael Collins "Carrying The Fire"*

THE "ORIGINAL 7"—Mercury Astronauts Wally Schirra, Deke Slayton, John Glenn, and Scott Carpenter. Top Row: Alan Shepard, Gus Grissom and Gordon Cooper.

By the Astronauts, as told to Dick Lattimer

Foreword by James A. Michener,
Author of the #1 Bestseller "SPACE"

A Mini-History of America's Manned Moon Program
Including America's Astronaut Patches and Callsigns.

Special thanks to the National Aeronautics and Space Administration who provided most of the photographs in this book and much of the information. America can be proud of NASA and its dedicated men and women of the 20th Century who are working to provide us with an even more exciting 21st Century.

Thanks also to the following people and newspapers who granted permission for use of their material in this book: Charles M. Schulz, Robert McCall, Harvey Publications, Inc., Frank Kelly Freas, Simon & Schuster, Inc., Little, Brown and Company, The South Bend Tribune, The Gainesville Sun, The Daily Oklahoman—July 21, 1969, Copyright 1969, The Oklahoma Publishing Co., The Detroit News, The News-Sentinel, Fort Wayne, Indiana, The Houston Post, The Indianapolis Star, The Wapakoneta Daily News, The Bay City Times, The Detroit Free Press, "Today" Brevard County, FL, The Miami Herald, The Boston Globe, The Denver Post Corp., Chicago Sun-Times, The Hammond, Indiana Times, The Chicago Tribune, The San Diego Union, The Philadelphia Inquirer, The Canton, Ohio Repository, Michael Collins—"Carrying the Fire", The Orlando Sentinel, The Washington D.C. Evening Star, Fort Wayne Journal-Gazette, The Cleveland Plain Dealer, The Atlanta Constitution, Parade Magazine, The Washington Post.

Design by C. M. Haller.

Library of Congress Registration No. TXU 116-653.
Fifth Printing

Printed in the United States of America.
ISBN 0-9611228-0-3

The Whispering Eagle Press
PMB 14
Post Office Box 147050
Gainesville, FL 32614-7050

DEDICATION

This edition is dedicated to all the people on Earth. You are the *"We"* in *"All We Did Was Fly To The Moon."* The Russian and American competition in the 1960s led to this magnificent accomplishment and it would not have been achieved without that "Space Race." Today many, many countries and their people are involved in the International Space Station as together we on this beautiful blue planet reach for the stars. When accomplished, that goal will insure the perpetuation of our species throughout the universe.

JAMES A. MICHENER *during research for his bestseller "Space."*
Photo Courtesy of NASA.

"You seem sometimes to forget that you and your glorious young men are engaged in an adventure that will command public interest for at least five centuries. You're not in America in 1965. You're in the books of World History in 2465. And if dedicated people like me do not tell your story accurately today, do you know how the books of 2465 are going to tell it?"

—Cynthia Rhee

An excerpt from "SPACE" by James A. Michener

FOREWORD

By James A. Michener

Each era of history progresses to a point at which it is eligible to wrestle with the great problem of that period. For the ancient Greeks it was the organization of society; for the Romans it was the organization of empire; for the medievalists the spelling out of their relationship to God; for the men of the fifteenth and sixteenth centuries the mastery of the oceans; and for us it is the determination of how mankind can live in harmony on this finite globe while establishing relationships to infinite space.

The life of any nation since the beginning of history has been a record of how it confronted the great challenges that inevitably came its way. History is a grand mix of concepts, actions, organizings and commitments which determines the extent to which any nation can achieve a good life for its citizens, and I believe without question that if a nation misses the great movements of its time it misses the foundations on which it can build for the future. The high technical requirements for success in space are so fundamental that spin-off rewards are almost automatic. Radio, television, medical instrumentation, minaturizing, watches, new food processes, communications, health advances and improvement in clothing are some of the few advantages which I myself have gained because of the space program, and I am speaking only of small items which can be comprehended and used by the individual.

If one considers the larger items, like intercontinental communications satellites, the mapping of weather patterns, the analysis of soils and forests, the exploration for minerals including oil, the management of fisheries and the like, the potential rewards are multiplied many times.

No one today can even guess the limits of either the personal items or the industrial which might accrue from the basic scientific work that has to be done in a space program. This is the great unknown ocean of the universe and we in 1983 are as obligated to probe it and use it and participate in its control as the nations of Europe were obligated to explore their terrestrial oceans in 1483.

I believe that there are moments in history when challenges occur of such a compelling nature that to miss them is to miss the whole meaning of an epoch. Space is such a challenge.

And among the hundreds of thousands of Americans who accepted this challenge in the early days of the Age of Space were the daring aviators who learned to fly in the airless reaches of the sky. This book is about them and about the symbols and insignias they chose to wear to represent both themselves and all of mankind. It is a fascinating slice of American History.

James A. Michener
Austin, Texas

Good old Yankee know-how built the incredible flying machine that took America to the moon. Here Rockwell International workers combine elbow grease and Alpha Waves on an early model of the Apollo Command Module.

"MAKING IT HAPPEN"

Thousands of Americans and American businesses were the keys to the success of our first Lunar Program. At the height of the Apollo Program in the 1960s, over 390,000 men and women in industry, 33,000 in NASA Centers and 10,000 in American universities were hard at work "making it happen." (See pages 136 through 142.)

The author's former employer, Kidde, Inc., Bellville, New Jersey, made the emergency flotation bag for the Mercury spacecraft; our Weber Aircraft Division in California made the ejection seats for the Gemini spacecraft and the foldable crew couches for the Apollo Command Module. As a matter of fact, our Weber ejection seat saved Neil Armstrong's life when he had to eject from the Lunar Landing Training Vehicle prior to the flight of Apollo II. During Project Mercury the author worked for the old Studebaker-Packard Corporation and our CTL Division (Cincinnati Testing Laboratory) made the Mercury spacecraft's heat shield. Nearly every American knows someone who played a key part in this remarkable chapter in American history.

Perhaps one of the most outstanding roles, however, continues to be that played by the men and women of Rockwell International. For over 25 years they have devoted almost their entire resources, 24 hours daily, to the design, development, fabrication and testing of America's Space Machines. They built all the Apollo spacecraft and service modules flown in space, including all the Lunar Missions covered in this book, the Apollos for Skylab and the Apollo-Soyuz Test Project. They also did the same for the powerful second stage of the Saturn launch vehicle used in Apollo. Today they continue their fine tradition by designing, developing, building and testing our entire fleet of STS Space Shuttle spacecraft. Perhaps their slogan says it all for their fine company and their dedicated staff, as well as for all America–"When A Spacecraft Lands On Earth . . . It Comes From Rockwell International."

ACKNOWLEDGEMENT

THE ASTRONAUTS

The author would like to thank the following Astronauts for their help in compiling this information: (in alphabetical order):

Neil Armstrong,
Gemini VIII, Apollo XI

Alan Bean,
Apollo XII, Skylab III

Vance Brand,
Apollo-Soyuz, Space Shuttle 5

Scott Carpenter,
Mercury (Aurora 7)

Jerry Carr,
Skylab III

Gene Cernan,
Gemini IX, Apollo X, Apollo XVII

Michael Collins,
Gemini X, Apollo XI

Pete Conrad,
Gemini V, Gemini XI, Apollo XII, Skylab I

Gordon Cooper,
Mercury (Faith 7), Gemini V

Walt Cunningham,
Apollo VII

Charlie Duke,
Apollo XVI

Joe Henry Engle,
Space Shuttle 2

Ron Evans,
Apollo XVII

Fred Haise,
Apollo XIII

Jim Irwin,
Apollo XV

Jack Lousma,
Skylab II, Space Shuttle 3

Jim Lovell,
Gemini VII, Gemini XII, Apollo VIII, Apollo XIII

T.K. Mattingly,
Apollo XVI, Space Shuttle 4

Jim McDivitt,
Gemini IV, Apollo IX

Ed Mitchell,
Apollo XIV

Stuart Roosa,
Apollo XIV

Wally Schirra,
Mercury (Sigma 7), Gemini VI, Apollo VII

Alan Shepard,
Mercury (Freedom 7), Apollo XIV

Deke Slayton,
Apollo-Soyuz

Tom Stafford,
Gemini VI, Gemini IX-A, Apollo X

Jack Swigert, Jr.,
Apollo XIII

Paul Weitz,
Skylab I

Al Worden,
Apollo XV

THE ASTRONAUT OFFICE

Cyril E. Baker, Houston

Nancy Gunter, Cape Canaveral

JOHNSON SPACE CENTER

John McLeish, Public Affairs
Mike Gentry, Public Affairs
Terry White, Public Affairs
Stan Jacobsen, Publications & Graphics
Jerry Elmore, NASA Artist
Barbara Matelski, NASA Artist
Joe Garino, Astronaut Trainer

NASA HEADQUARTERS

Dr. Monte Wright, Director, History Office
Lee Saegesser, NASA Archivist
Eleanor Ritchie, NASA Writer
Carrie Karegeannes, NASA Editor

The author would also like to thank the following people and companies for providing information for this historical record:

Don Fuqua, Chairman
Committee on Science & Technology
U.S. House of Representatives
Washington, D.C.

Roseann Smith
Grumman Aerospace Corp.
Bethpage, N.Y.
(Lunar Module)

Billy Lapham
ILC Industries
Wilmington, Delaware
(Apollo Spacesuits)

Evan McCollum
Martin Marietta Aerospace
Denver, Colorado
(Gemini Rocket, Skylab, Space Shuttle)

Frank B. Thomas
U.S. Postal Service
Washington, D.C.
(Space Stamps)

N. Gordon LeBert
McDonnell Douglas Corp.
St. Louis, Missouri
(Mercury, Gemini Capsules)

Dale A. Jones
Director of Wildlife & Fisheries
U.S. Forest Service
Washington, D.C.

Sue Cometa
Rockwell International
Downey, California
(Apollo Command Module, Saturn V 2nd Stage)

Bill Rice
Boeing Aerospace Company
Seattle, Washington
(Saturn V Rocket)

Ernst Wildi
Victor Hasselblad, Inc.
Fairfield, N.J.
(Crew Cameras)

Fred K. Godfrey
Weber Aircraft
KIDDE, Inc.
Burbank, California
(Gemini Escape System, Apollo Couches, LLTV Escape System)

John H.K. Saxon
Honeysuckle Creek Tracking Station
Australian Capital Territory
Spacecraft Tracking & Data Network
Canberra, Australia

May 25, 1961—President John F. Kennedy urges a National Goal of "landing a man on the moon and returning him safely to Earth," in an "Urgent National Needs" speech to a joint session of Congress.

THE CHALLENGE

"First, I believe that this nation should commit itself to achieving the goal, before this decade is out, of landing a man on the moon and returning him safely to the earth. No single space project in this period will be more impressive to mankind, or more important for the long-range exploration of space; and none will be so difficult or expensive to accomplish. We propose to accelerate development of the appropriate lunar space craft. We propose to develop alternate liquid and solid fuel boosters, much larger than any now being developed, until certain which is superior. We propose additional funds for other engine development and for unmanned explorations—explorations which are particularly important for one purpose which this nation will never overlook: the survival of man who first makes this daring flight. But in a very real sense, it will not be one man going to the moon if we make this judgment affirmatively, it will be an entire nation. For all of us must work to put him there.

"Secondly, an additional 23 million dollars together with 7 million dollars already available, to accelerate development of the NOVA nuclear rocket. This gives promise of some day providing a means for even more exciting and ambitious exploration of space, perhaps beyond the moon, perhaps to the very end of the solar system itself.

"Third, an additional 50 million dollars will make the most of our present leadership, by accelerating the use of space satellites for world-wide communications.

"Fourth, an additional 75 million dollars—of which 53 million dollars is for the Weather Bureau—will help give us at the earliest possible time a satellite system for world-wide observation.

"...If we are to go only half way, or reduce our sights in the face of difficulty, in my judgement it would be better not to go at all.

"It is a most important decision that we make as a nation. But all of you have lived through the last four years and have seen the significance of space and the adventures in space, and no one can predict with certainty what the ultimate meaning will be of mastery of space.

"I believe we should go to the moon."

John Fitzgerald Kennedy
Washington, D.C.
May 25, 1961

Table of Contents

"ALL WE DID WAS FLY TO THE MOON"

By the Astronauts, as told to Dick Lattimer

The author and his family at the Kennedy Space Center, July 16, 1969, the day Apollo 11 left to carry the first fire to the Moon. Back row, Dick and Alice Lattimer. Front row, left to right: Elizabeth, Scott and Michael Lattimer. The day marked Beth's 8th birthday.

INTRODUCTION

Welcome...

This history was written primarily by the men who made our first flights into outer space. These pioneers have set down their memories of exactly why they christened their spaceships as they did and the symbolism that they chose to wear on their spacesuits. You and I owe them our thanks for sharing this colorful bit of Americana with us.

Other invaluable sources included Dora Jane Hamblin's "Spacecraft Anonymous", Life Magazine, Vol. 65, No. 15, 1968; "Carrying the Fire" by Michael Collins, Ballantine Books; "Return To Earth" by Edwin Aldrin, Jr., Random House; "Gemini" by Gus Grissom, Macmillan Company; "We Seven" by the Original 7 Astronauts, Simon & Schuster; "First on the Moon," Little Brown & Company; Analog Science Fiction Magazine, June, 1973; and the histories, mission reports and press releases of the National Aeronautics & Space Administration.

The author assumes all responsibility for any errors and thanks all who have helped assemble this historical record, especially the Astronauts themselves. Also my good friend Cy Baker, along with Barbara Matelski, Stan Jacobsen, Mike Gentry and John MacLeish at the Johnson Spacecraft Center in Houston.

Having once been a struggling young writer himself working on his first book, Pulitzer-Prize author James A. Michener knows the thrill it is for me to have his Foreword introduce this bit of contemporary American history. I am very grateful for his help.

And, finally, my thanks to my wife, Alice, and my children Mike, Beth and Scott. We stood together in the hot summer sun early one morning and watched them carry the first fire to the moon.

Dick Lattimer

Dick Lattimer
Cedar Key, Florida

THE NASA INSIGNIA

The Original 7 Mercury Astronauts wore the NASA Insignia on their spacesuits and nothing else. They were permitted to personalize their flights by naming their space "capsules" as they were referred to in those days.

The NASA Insignia was designed by the Army Institute of Heraldry and approved by the Commission on Fine Arts and the NASA Administrator. It has a dark blue sky background; solid red wing configuration; white inner elliptical flight path, stars and the letters NASA standing for the National Aeronautics and Space Administration.

Left to right: Lt. M. Scott Carpenter, USN; Capt. L. Gordon Cooper, Jr., USAF; Lt. Col. John H. Glenn, Jr., USMC; Capt. Virgil I. "Gus" Grissom, USAF; Lt. Comdr. Walter M. Schirra, Jr., USN; Lt. Comdr. Alan B. Shepard, Jr., USN; and Capt. Donald K. "Deke" Slayton, USAF.

PROJECT MERCURY

"The fictional literature of space travel dates at least from the second century A.D. Around the year 160 the Greek savant Lucian of Samosata wrote satirically about an imaginary journey to the Moon, "a great countrie in the aire, like to a shining island." Carried to the Moon by a giant waterspout, Menippus, Lucian's hero, returns to Earth in an equally distinctive manner: The angry gods simply have Mercury take hold of his right ear and deposit him on the ground.

"Abe Silverstein (The Lewis Aeronautical Laboratory) advocated a systemic name with allegorical overtones and neutral underpinning; The Olympian messenger Mercury...was the most familiar of the Gods in the Greek pantheon to Americans. Mercury, the son of Zeus and grandson of Atlas, with his winged sandals and helmut and caduceus, was too rich in symbolic associations to be denied. On Wright Brothers Day, December 17, 1958, 55 years after the famous flights at Kitty Hawk, North Carolina, NASA announced in Washington that the manned satellite program would be called "Project Mercury." –NASA

This New Ocean
A History of Project Mercury

Shown is the Mercury monument at Launch Pad #14, John F. Kennedy Space center. There was no project patch for Mercury as there was for subsequent projects. This Monument shows the astronomical symbol for the Planet Mercury and the number 7 honoring the "Original 7" Astronauts.

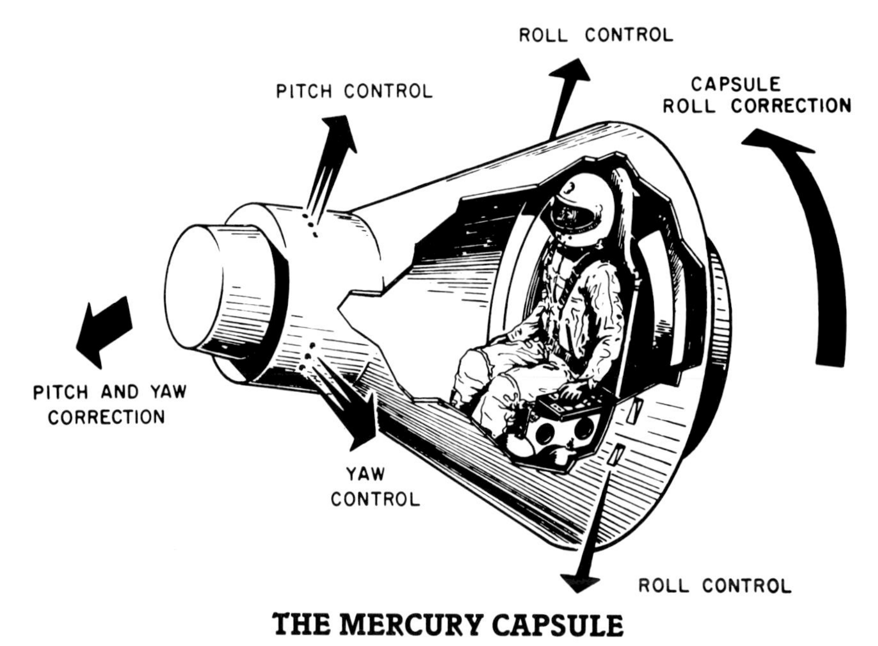

THE MERCURY CAPSULE

The size of our first space capsule was dictated by the proven rockets of the day. Outside dimensions were 9 feet 6 inches tall and 6 feet 2 inches in diameter at its widest spot. Yet the interior pressure hull was about the size of a telephone booth. John Glenn has said, "You don't climb into the Mercury spacecraft, you put it on." When seated on his contour couch inside the capsule the instrument panel was only 24 inches in front of the Astronaut's face. Mercury would fit on the top of both the Redstone and Atlas boosters. It was built by McDonnell Aircraft in St. Louis with the help of 4000 suppliers, including 596 direct subcontractors from 25 states. Mercury contained 750,000 parts and 7 miles of wiring.

Before entrusting a man to Mercury we flew monkeys, chimpanzees and even a pig to test out the capsule and systems.

A major feature of the bell-shaped Mercury capsule was an ablative heat shield made by the CTL division of The Studebaker Corporation where the author worked at the time. This absorbed the tremendous heat built-up on reentry into the atmosphere.

THE MERCURY ROCKETS

Two boosters were used during the Mercury flights: the Chrysler-built Redstone for the Shepard and Grissom suborbital flights and the General Dynamics Convair Atlas used on the remaining Mercury flights. Atlas was 10 feet in diameter and 71 feet tall. The Mercury escape tower was 16 feet tall and featured a solid-propellent rocket to pull the capsule away from the booster rocket in case of trouble during launch. It never had to be used.

HOME EDITION

The South Bend Tribune.

U. S. PLANS EARTH SATELLITE

Local 5 Members Move to Stay on Jobs

IKE APPROVES PROJECT FOR '57 LAUNCHING

GOP CAN STAY IN FOREVER, IKE DECLARES

Dying Mother Takes Child On 'Vacation'

SHAKY UNITY DISPLAYED IN MASS VOTING

Lawmakers Rush Toward Adjournment

Chicago Daily Tribune

FINAL

RED IN SPACE SEES EARTH!

U.S. Indicts General Motors

ABLE TO DISTINGUISH PEAKS, CITIES, OCEANS

EVIL GENIUS TAG PINNED ON EICHMANN

CHARGE NEAR MONOPOLY IN RAIL DIESELS

State Asked by City: Ban Housing Bias

On July 29, 1955 President Eisenhower announced plans for the launching of a basketball-sized U.S. satellite to be orbited in "late 1957." Most of America had forgotten that in the shock of Russia's launch of the world's first satellite, Sputnik, in October 1957. We had conveniently provided the Soviets with a target date to beat, and beat us they did. Somewhat the same situation existed with the first manned flight. Yuri Gagarin blasted off on his historic first flight even while Alan Shepard's Mercury-Redstone rocket sat on its launch pad No. 5 waiting for final adjustments to his space capsule. The Soviets had beaten the Yanks to the punch one more time. Such was the price of a free and open society, yet the Race to the Moon had just begun.

A. Escape rockets, tower jettison rockets, and escape tower provide safe recovery of vehicle in case of booster malfunction.

B. Antenna housing for ground command, telemetry and voice antennas; six-foot drogue parachute; and infrared horizon scanners for attitude reference.

C. Recovery compartment contains the 63-foot diameter main and reserve parachutes; recovery beacon antennas; flashing recovery-aid light.

D. Crew compartment contains major spacecraft systems, including communications, electrical power, environmental control, instrumentation, navigation aids, stabilization and control.

E. Retrograde package contains three retrograde rockets for initiating the spacecraft's return from orbit; and three rockets for separating the spacecraft from the booster after orbital velocity is reached.

F. Heat shield provides protection for the astronaut from the extreme temperatures experienced during re-entry.

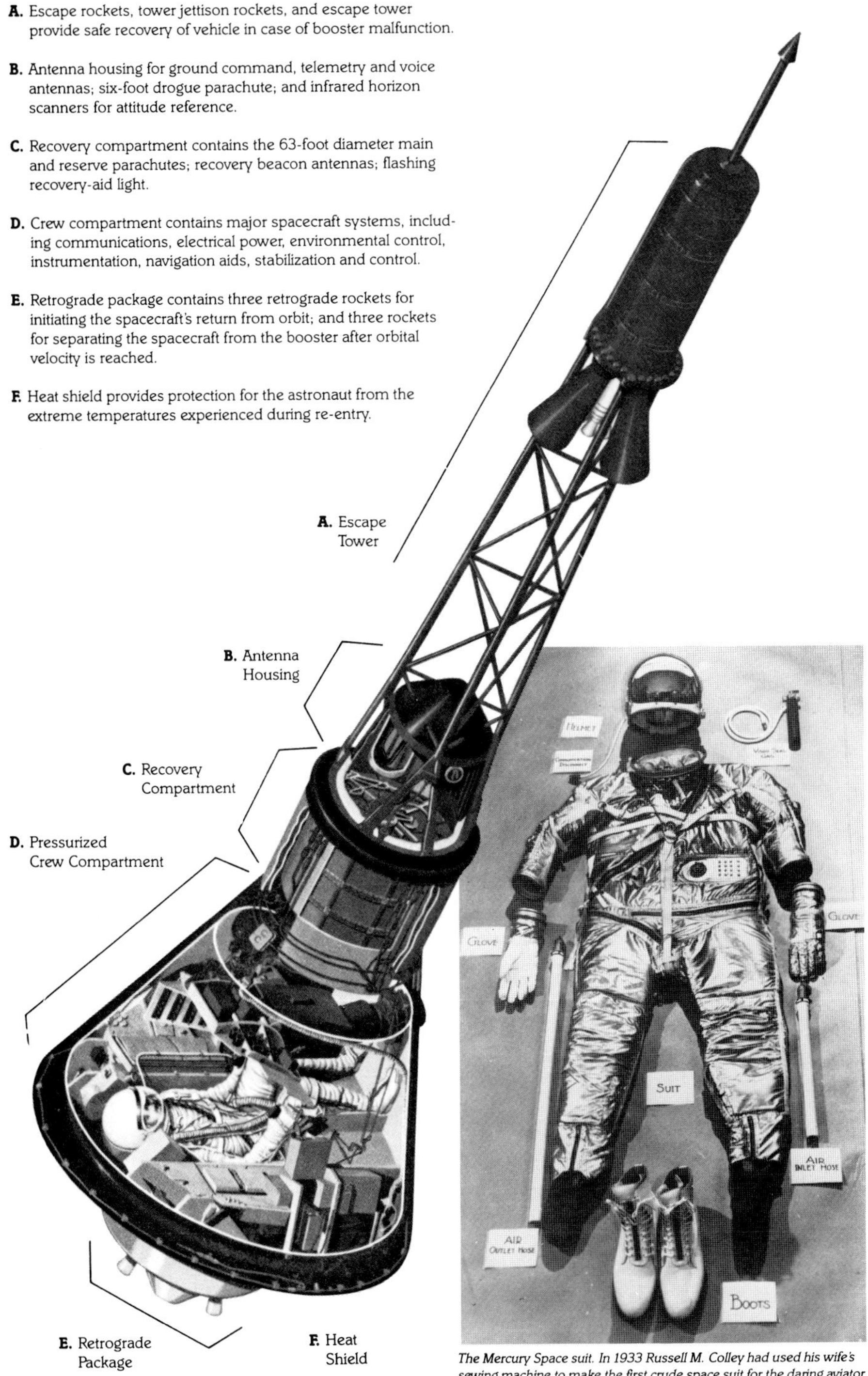

The Mercury Space suit. In 1933 Russell M. Colley had used his wife's sewing machine to make the first crude space suit for the daring aviator, Wiley Post, to use on a high altitude flight in his famous Winnie Mae. Colley, Carl F. Effler, D. Ewing, Astronaut Wally Schirra and others at the B.F. Goodrich Company in Akron, Ohio and in NASA developed this spacesuit for Project Mercury. It is not known whether Colley again used his wife's sewing machine.

Alan B. Shepard, Jr. in Freedom 7 just prior to its being sealed on the morning of launch. The "Man in the Can" label seems quite evident in these cramped quarters.

Al Shepard inspecting Freedom 7 aboard the U.S. Navy Carrier Champlain following his historic suborbital spaceflight.

Shepard being hoisted aboard the recovery helicopter.

HOME EDITION

The South Bend Tribune

ASTRONAUT SHOT DELAYED

Talks Quiet Laos Guns

CLOUDY SKY POSTPONES SPACE TRY

Shepard Has Faced Many Challenges

U.S. AIRLINER FORCED TO FLY TO CUBA

Hopes Rise For End of Civil War

America did not know who its First Astronaut was to be until the launch was "scrubbed" due to cloud cover on Tuesday, May 2, 1961. The identity of Alan B. Shepard, Jr. was revealed only after the countdown had been cancelled, 2 hours and 20 minutes before launch. This was in case a last minute substitution of his back-up pilot, John Glenn, became necessary. Yuri Gagarin of the Soviet Union had become the first man in space just 20 days earlier aboard his small Vostok 1 capsule, codenamed Cedar.

HOME EDITION

The South Bend Tribune

SPACEMAN AWAITS SHOT

Wage Bill Aids JFK Program

Begin Counting For 7 a.m. Flight

POLICE FIND BOY SOUGHT IN SLAYINGS

Holland Youth, 16, Picked Up in Dakota

TREMENDOUS BOOST GIVEN BROAD PLANS

LABOR BLAMED IN SPACE LAG

'Myth of the Seas' Sinks School Yacht

The launch was again delayed on Thursday because of weather. A small item on this front page announced that even though America was behind in the Space Race, the "U.S. Plans to Be First on Moon. Dr. Robert C. Seamans, Jr., associate administrator of NASA, said yesterday this nation hoped to be hard at work charting the moon's surface when Soviet explorers arrived. A round trip flight to the moon is scheduled for 1967, according to Seamans." (UPI).

FREEDOM 7

Alan B. Shepard, Jr.
East Derry, New Hampshire

Series: Mercury
Date: May 5, 1961
Crew & Age: Alan B. Shepard (37)
Distinction: First American in Space (suborbital)
Flight Duration: 15 minutes, 22 seconds
Crew Insignia: NASA Patch
Spacecraft Name: "Freedom 7"

"Pilots have always named their planes. It's a tradition. It never occurred to me not to name the capsule. I checked with Dr. Gilruth and I talked it over with my wife and with John Glenn, who was my backup pilot. We all liked it."

—Alan B. Shepard

Shepard added the "7" because the capsule he flew was factory Model No. 7. Everybody thought it represented the Original Seven Astronauts, and that seemed such a good idea at the time that all of the rest of the Mercury series names also carried the "7".

FREEDOM 7

HOME EDITION

The South Bend Tribune.

SPACE FLIGHT A SUCCESS

Shepard Lands Safely After 5,100 m.p.h. Trip

RED ABUSE THREATENS LAOS TALKS

Balloonist Sets Mark, But Drowns

MASS TERROR REPORTS LEAK OUT OF CUBA

JFK Lauds Astronaut Over Phone

U.S. PARTLY EVENS SCORE WITH RUSSIA

Wearing a parachute in case of an emergency, Shepard lifted off from Pad 5 at Cape Canaveral perched on top of a slim black-and-white Redstone rocket at 9:34 a.m. on May 5, 1961. The flight lasted 15 minutes and 22 seconds, reached an altitude of 116.5 miles at a maximum speed of 5180 miles per hour and landed 302 miles downrange. Space officially begins at 100 miles altitude. Shepard might have beaten Gagarin into space except for a faulty electrical relay on the Mercury capsule used to shoot "Ham the Chimpanzee" into space causing him to overshoot his Atlantic Ocean target area. Another unmanned test had to be scheduled before we would commit an American to space and Russia's Gagarin beat Shepard to the honor.

HOME EDITION

The South Bend Tribune.

SHEPARD GIVES U.S. BIG LIFT

Rebels Break Laos Peace

FREEDOM 7 MAY BE PUT IN MUSEUM

Yanks Cheered By Achievement

RED TROOPS ATTACK LOYAL GUERRILLAS

Dispute Puts Parking Up To Council

Al Shepard laid his life on the line for his country and gained a measure of immortality. Few remember that just 11 days before his flight an attempt to orbit a "mechanical astronaut" aboard an Atlas Rocket had been destroyed just 40 seconds after lift-off. The Chrysler Corporation built the Redstone rocket that put the U.S. into space and McDonnell Aircraft Corporation built the bell-shaped Mercury "capsule." When he entered the spacecraft that morning Shepard found a note taped to the instrument panel by his back-up John Glenn. It read "No Handball playing in this Area." The small Freedom 7 capsule was just 74 inches wide.

Liberty Bell 7 is mated to its Redstone rocket on the launch pad.

Back-up pilot John Glenn helps Gus Grissom wiggle into his Liberty Bell 7 spacecraft on launch morning, July 21, 1961.

Liberty Bell 7 sank moments after this picture was taken following Grissom's successful suborbital space flight.

HOME EDITION

The South Bend Tribune.

REJECT RUSS BERLIN STAND

Space Shot Set for 8 A.M.

Reds Girdle City With Military Ring

NOTES CLAIM SOVIET VIEW PERILS PEACE

WEATHERMEN SEE BREAK BY TOMORROW

RESCUE CREW PLANS EERIE CAVE SEARCH

HOME EDITION

The South Bend Tribune.

NATION TO HEAR JFK PLANS

Tuesday Talk to Reveal Defense Measures

CAVE YIELDS BODIES OF 2 I.U. STUDENTS

Plane Crash Kills Wife of Aide to JFK

SPACE SHOT RESCHEDULED FOR FRIDAY

Reply on Berlin Sternly Worded

TUNISIA FIRES UPON FRENCH

The launch of Liberty Bell 7 was originally scheduled for July 18, but weather forced a postponement. In Berlin the "Cold War" was heating up and in Washington the President, John F. Kennedy, had just gone before a Joint Session of the U.S. Congress and made his famous May 25 speech—"I believe that this nation should commit itself to achieving the goal, before this decade is out, of landing a man on the Moon and returning him safely to Earth." America had not yet orbited a man in space. "Man's reach should always exceed his grasp."

Another "scrub" occurred on Wednesday, July 19, after Gus had sat in his capsule on top of his Redstone rocket for 3 hours and 22 minutes. In Berlin the Communists had 67,500 troops and 1,200 tanks surrounding the beleaguered city. And Russian premier Nikita Krushchev had called Alan Shepard's Freedom 7 suborbital flight "a flea jump" compared to the USSR orbiting the Earth and sending aloft a 7 ton satellite. America's pride was smarting.

LIBERTY BELL 7

Virgil I. "Gus" Grissom
Mitchell, Indiana

Series: Mercury
Date: July 21, 1961
Crew & Age: Virgil I. "Gus" Grissom (35)
Crew Insignia: NASA Patch
Spacecraft Name: "Liberty Bell 7"
Flight Duration: 16 Minutes

Gus Grissom chose the name Liberty Bell because the capsule was shaped like a bell and because it carried stirring American connotations so important during the early days of the American space program.

"John Glenn felt that the symbolic number "7" should appear on all our capsules in honor of the team, so this was added," Gus stated in the Mercury Astronaut's book "We Seven."

"Then one of the engineers got the bright idea that we ought to dress Liberty Bell up by painting a crack on it just like the crack on the real one. No one seemed quite sure what the crack looked like, so we copied it from the "tails" side of a fifty-cent piece. Ever since my flight, which ended up with the capsule sinking to the bottom of the Atlantic, there has been a joke around the Cape that that was the last capsule we would ever launch with a crack in it."

—Gus Grissom

LIBERTY
BELL
7

HOME EDITION

The South Bend Tribune.

GRISSOM BACK; CONE LOST

French Pound Bizerte Canal

U. S. Astronaut Picked From Sea

FULL ATTACK LAUNCHED TO REGAIN PORT

Rail Tunnel Burns; Lets House Fall In

FLIER TALKS TO PRESIDENT

The suborbital flight finally took place on Friday, July 21. Gus won everyone's hearts when he admitted that he had been "a bit scared" at liftoff. But that and the actual flight went smoothly. He did, however, nearly drown when the capsule hatch suddenly blew off after splash-down and sea water began to pour into the Liberty Bell 7. He barely managed to escape, but Liberty Bell 7 sank in 16,800 feet of sea water. He then began to worry about sharks, but was finally rescued. His helmet floated out of his sinking capsule and they found it next to a 15 foot shark, his fears had been well-founded.

108 PAGES

The South Bend Tribune.

TWENTY CENTS

TUNISIAN FIGHTING STOPS

Hatch Shot Accidental: Grissom

Heed U.N. Call For Cease-Fire

Blames Malfunction For Loss of Capsule

Star Costs Gus 2 Steaks Dinners

Liberty Bell 7 still lies at the bottom of the Atlantic Ocean 303 miles southeast of Cape Canaveral. Its protective cocoon had taken Gus Grissom to an altitude of 118 miles and a speed of 5,280 miles an hour. A small news item this day read: "One of the first things astronaut Gus Grissom did in his space flight was to lose a bet. He had wagered with his two fellow astronauts that he would not see a star on his daytime flight. As soon as he looked out his window in the capsule, Grissom said, "Right there in the middle was a single star. I knew right then I had lost two steak dinners."

John Glenn entering Friendship 7 during pre-flight rehearsals.

Lift-off of John Glenn and Friendship 7.

Close-up of the notebook John Glenn carried on Friendship 7.

HOME EDITION. The South Bend Tribune.

U.S. NAMES 2 SPACE TEAMS

U.N. Group Votes Nuclear Curbs

Second Orbit Now in Planning

Rocky Tells Wife About Hunt for Son

ASSEMBLY PASSAGE IS CERTAIN

U.S. and Britain Oppose Arms Limits

USAF Records ICBM Successes

RED WALKS OUT ON AWARD FETE

Kennedy May Visit Two Latin Nations

"Heartened by the world-circling space flight of a chimpanzee, the United States picked two space teams to take the next steps up the long, dark stairway to the moon." So read the Associated Press story announcing the selection of John Glenn for the first U.S. orbital flight with Scott Carpenter as his back-up. Donald K. Slayton would make the second orbital flight backed up by Walter Schirra, Jr., or so the initial plans went.

HOME EDITION. The South Bend Tribune.

CLOUDS STOP GLENN FLIGHT

U.S. Gives Ground on Cuba

MOON PHOTOS EXPECTED FROM ROBOT

Expect Delay Until Tuesday

DELAY SEEN FOR REMOVAL FROM OAS

Barges Rip Moorings, Ram Piers

Originally scheduled for January 23, the launch next slipped to January 27 due to bad weather. As this headline shows, that date, too, was missed after Glenn had sat in Friendship 7 a little over 5 hours patiently waiting to be launched. America's 727 pound Ranger III unmanned spacecraft was halfway to the Moon on a photography assignment.

FRIENDSHIP 7

John H. Glenn, Jr.
Cambridge, Ohio.

Series: Mercury
Date: February 20, 1962
Crew & Age: John H. Glenn, Jr. (40)
Distinction: First American to Orbit the Earth
Crew Insignia: NASA Patch
Spacecraft Callsigns: "Friendship 7"
Flight Duration: 3 Orbits

"I put my wife Annie and the kids to work studying dictionaries and a Theasaurus to come up with a list of suitable names on which the family could vote. We played around with Liberty, Independence, a lot of them. The more I thought about it, the more I leaned toward the name Friendship. Flying around the world, over all those countries, that was the message I wanted to convey. In the end that was the name the kids liked best, too. I was real proud of them."

—John Glenn

"I am an Eagle" radioed Russian Cosmonaut Gherman Titov during a 17 orbit spaceflight of Vostok 2 on August 6, 1961, just days after Gus Grissom's flight. Titov traveled 435,000 miles, nearly far enough to go to the Moon and back. Nikita Krushchev further rubbed salt in America's wound by proclaiming that "Titov's feat has shown once again what Soviet man, educated by the Communist Party, is able to do." Five days after Titov's flight Russia closed East Berlin and began building the Berlin Wall.

HOME EDITION

The South Bend Tribune.

GLENN SCORES U.S. TRIUMPH

Marine First American to Orbit Earth

JFK URGES PAY RAISE FOR WORKERS

JOY SHOWN BY FAMILY, NEIGHBORS

Wife and Children Watch TV at Home

Flier Declares Condition Good

Algerian Rebellion Moves to Climax

CROWDS PRAY FOR SUCCESS

The stormy Florida February weather plus a fuel leak caused another 3 week delay in Glenn's flight but finally on the morning of February 20 at 2:20 A.M. John Glenn was awakened, for a planned 7:30 lift-off. But again the weather and then a booster guidance problem caused delays. Finally, however, at 9:47 A.M. Friendship 7 left Earth. A faulty heat shield signal indicating that the Astronaut would be incinerated upon reentry caused grave concern to Mercury Control and an apprehensive nation. Finally after a 4 hour and 55 minute three orbital flight Glenn landed safely in the Atlantic but not before Glenn experienced his worst emotional stress of the flight as burning chunks of material went flying by his window. Thinking that it might be his heatshield disintegrating he could do nothing but continue with reentry. He had earlier radioed, "That's a real fireball outside." Glenn earned a month's flight pay, $245, for his flight.

HOME EDITION

The South Bend Tribune.

GLENN IN EXCELLENT SHAPE

Holovachka Found Guilty

TAX EVASION INDICTMENT IS UPHELD

NIKITA ASKS U.S. TO POOL SPACE WORK

Tells Experts About Flight

Capital Wheels Stop As Rocket Blasts Off

Glenn had lost five pounds, five ounces during his flight and had not suffered the same space sickness that Titov had reported on his Soviet flight. However, in detonating his hatch in order to leave Friendship 7 aboard the Destroyer Noa, he did cut his knuckles slightly. He had been lifted aboard the recovery ship while still in the spacecraft. We did not want a repeat of Gus Grissom's close call on our first orbital flight.

Cecelia Bibby paints the name on Aurora 7.

Scott Carpenter in the Mercury simulator at Langley Field, Virginia. Mercury was no place for people with claustrophobia.

Carpenter looks into Aurora 7 prior to insertion. At the end of the flight he would have to crawl through the narrow top of the capsule like a man coming out of a bottle so that the capsule wouldn't sink.

Carpenter is finally found and is shown here being lifted aboard the Navy recovery helicopter.

HOME EDITION

The South Bend Tribune.

MARINES LAND IN THAILAND

Weather May Delay Orbit

LAUNCHING ODDS SET AT 50-50

Plane Crash Kills Africa Recovery Team

U.S. WEIGHS CONTINUING LAOTIAN PAY

Prepared to Stay 'Long as Needed'

Find Couple Shot Near

WHITE HOUSE PETS CLEARED

Estes Case to Get Full-Scale Airing

HOME EDITION

The South Bend Tribune.

CARPENTER FOUND ON RAFT

Plane Sights Space Capsule Floating on Ocean

SENATE AIDES QUIZ 1,000 IN ESTES CASE

Salan Term Is 'Triumph Of Terror'

FBI Enters Probe Into Death of Marshall

NATION SEES LIFTOFF ON TELEVISION

Astronaut's Wife Watches From Beach

Craft 200 Miles Beyond Ships

"I am an archer" wrote Scott Carpenter on his application to become an Astronaut. And like a bowhunter who faces his challenge alone with just his bow and arrow, Scott rode his Mercury capsule alone on a saga mixed with the poetry of sunset in space and the fear of instant death. The fourth American spaceflight was to have been Deke Slayton's. But after training for it for 10,000 hours, he was grounded because of a heart murmur. On March 15 he was replaced by Carpenter, a Navy test pilot from Colorado. Weather again forced delays in lift-off.

Millions of Americans sat down to drink their morning coffee and watch Aurora 7 lift-off at 7:45 A.M. As he passed over the Cape after one orbit he deployed a multicolored balloon to study the effects of space on the reflection properties of colored surfaces, and to obtain aerodynamic drag measurements. He used more fuel than planned and drifted for most of his 3rd and final orbit. Then things began to sour. He overshot his landing site by 250 miles and an anxious America held its breath for 55 long minutes until he was found bobbing in his liferaft 1,000 miles southeast of the Cape.

AURORA 7

M. Scott Carpenter
Boulder, Colorado

Series: Mercury
Date: May 24, 1962
Crew & Age: Malcom Scott Carpenter (37)
Distinction: First Astronaut to eat food in space.
Crew Insignia: NASA Patch
Spacecraft Name: "Aurora 7"
Flight Duration: 3 orbits

"I considered naming my capsule Rampart 7 after the Colorado mountain chain. That probably would have been a better name than Aurora. It's shorter and more positive. It would have come through the static better. But I liked Aurora 7. It has a celestial significance and it had a sentimental meaning to me because my address as a child back in Colorado was on the corner of Aurora and 7th Streets in Boulder."

—Scott Carpenter

Cecelia Bibby, Chrysler Corporation employee, paints the name Aurora 7 on Scott Carpenter's spacecraft while he looks on.

Flight Story in Pictures--Page 11 and Back Page

Chicago Daily Tribune

FINAL

SAVE ASTRONAUT AT SEA

Overshoots Target 250 Miles After 3 Orbits

CITY GRIPPED BY SILENCE OF AURORA 7

Then Comes the Cry of Joy

Capsule Soars Into Orbit—Family Proud

CARPENTER RESCUED AFTER 3 HOURS IN SEA

U. S.' 2d Spaceman Floats in Raft, 'Feeling Fine,' When Copters Arrive

Scott had radioed back from space that "The sunsets are most spectacular. The Earth is black after sun has set... The first band close to the Earth is red, the next is yellow, the next is blue, the next is green, and the next is sort of a... sort of a purple. It's almost like a very brilliant rainbow." He spent 3 hours in his raft before being picked up by helicopter, part of the time with two rescue frogmen—John F. Heitsch and Ray McClure. Carpenter had wiggled out of the narrow top of the Mercury capsule rather than blow-off the hatch and risk the capsule's sinking.

HOME EDITION

The South Bend Tribune

NOT CONFUSED: CARPENTER

CAB Pinpoints Blast on Jetliner

Tells of Tasks On Third Orbit

Crews Reconstruct Wreckage of Plane

TEXAN HINTS AT NEW ANGLE IN ESTES CASE

State Official Says Clement Paid

STOCK PRICES LOWER AGAIN

As Carpenter began aligning Aurora 7 for reentry, he suddenly found himself in trouble. NASA's history of the flight indicates that "the automatic stabilization system would not hold the 34-degree pitch and zero-degree yaw attitude. Aurora 7 was actually canted at retrofire about 25 degrees to the right." This and other factors contributed to the 250-mile overshot. But Scott had to take the "heat" as this headline suggests.

Sigma 7 is hoisted into position at the launch pad.

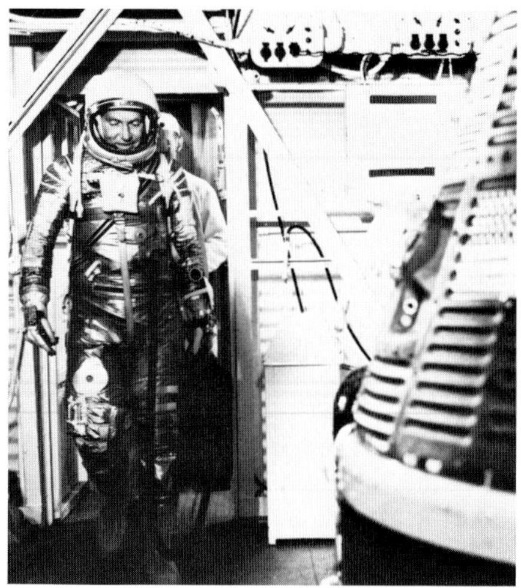

Wally Schirra strides confidently toward his Sigma 7 spacecraft on launch day.

Schirra is assisted by back-up pilot L. Gordon Cooper and NASA engineers into Sigma 7 prior to launch at 7:15 AM, October 3, 1962.

A Navy whaleboat from the Carrier Kearsarge towed Schirra and Sigma 7 to the waiting ship.

HOME EDITION

The South Bend Tribune

ORBIT MARATHON EXTENDED

Tandem Feat Puts Pressure on Project Gemini

2 SPACEMEN ARE SYMBOLS OF DREAMS

World Talks Of Lunar Goal Next

U.S. LAUNCH OF 2 MEN STILL LAGS

Cosmonauts Pass Reentry Position

POLITBUREAU East-West Battle

As Wally Schirra trained for his Sigma 7 flight, NASA received another jolt when the Soviet Union sent aloft a new 5-ton Vostok spacecraft, Falcon, on August 11 with Andrian Nikolayev aboard. Pavel Popovich aboard Vostok IV, Golden Eagle, rocketed up the following day. They came as close as 3 miles together and many expected an orbital rendezvous. It would not be until later that we would learn that the Russians could not master the secret of rendezvous. At the time, though, the first dual mission did little for our morale. Nikolayev completed 64 orbits, Popvich 48 orbits.

HOME EDITION

The South Bend Tribune

SCHIRRA TRIES SIX ORBITS

Two-Ton Capsule Making Perfect Flight

ASTRONAUT EASYGOING AND CALM

EXTENSION PHONE SET IN HELMET

Flier Corrects Problem of Heat

SEEK RELEASE FOR WALKER

Hangings in Effigy Upset Campus Calm

WIFE VIEWS ROCKET SHOT

Finally, on October 3, Wally Schirra, the Dwight Morrow High School graduate from Hackensack, New Jersey quipped to a friend, "I've got nothing better to do today, so I guess I'll take a trip around the world." America had chalked up 10 hours and 22 minutes in space, the Russians 192 hours and 25 minutes as Wally's Atlas rocket rose off the Cape's Pad 14. His escape tower "really said sayonara" according to this veteran of 90 combat missions in Korea. Problems with his spacesuit almost brought him back to Earth after only one orbit, but he was able to solve the problem and sail on. At one point, while the automatic pilot was on, he radioed down "I'm in chimp configuration", and a little bit later, "I guess that what I'm doing right now is sort of a couple of Immelmanns across the United States." His father had been a WW I flying ace and his mother a daring wing-walker. At reentry after 6 orbits he said the "Bear" he rode felt "as stable as an airplane."

SIGMA 7

Walter M. Schirra, Jr.
Hackensack, New Jersey

Series: Mercury
Date: October 3, 1962
Crew & Age: Walter M. Schirra, Jr. (39)
Distinction: Engineering Flight
Crew Insignia: NASA Patch
Spacecraft Name: "Sigma 7"
Flight Duration: 6 orbits

"Sigma means 'sum of'...a mathematical term. I wanted to get off the "Gee Whiz" names and use a technical/test pilot term as well as acknowledge the Original 7.

"I also toyed for awhile with the names Phoenix and Pioneer, but settled on Sigma because the flight was the sum of the efforts and energies of a lot of people."

—Wally Schirra

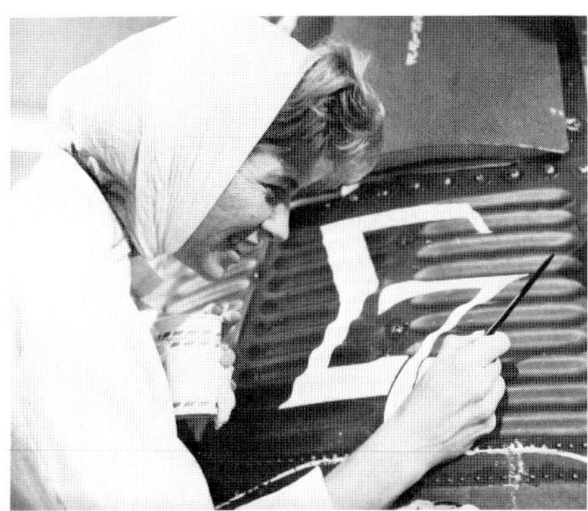

Cece Bibby painting the Sigma 7 design on Wally Schirra's Mercury Capsule.

Chicago Daily Tribune

FINAL

TV SENT BY NEW SATELLITE

How Trucker Paid Off Cops

U.S. AND EUROPE SEE PICTURES FROM SPACE

Veeck Sizes Up Phil Wrigley, a Diamond Riddle

A Business Genius, Who Doesn't Give a Hang for Baseball, Is the Candid Verdict

DOLED OUT $10 BILLS, WHISKY AT CHRISTMAS

Balk at Lie Test, 2 Aids Fired

A new era in communications was ushered in on July 10, 1962 when America put the Telstar satellite into orbit and transoceanic television became a reality. The satellite amplified the picture 10 billion times before retransmitting it. Unexpectedly, the first pictures were picked up in both England and France.

HOME EDITION

The South Bend Tribune.

X-15 CLIMBS TO 58 MILES

Rail Merger to Boost Elkhart

White's Flight Sets New Mark

YARD TRAFFIC IS EXPECTED TO DOUBLE

HEALTH CARE VOTE NEAR IN SENATE

Veteran test pilot Robert M. White became the first to qualify as an Astronaut in a winged aircraft when he flew the X-15 rocket plane to an altitude of 310,000 feet (58 miles) at a speed of 3,784 m.p.h. during his climb. One must fly to an altitude of 50 miles to earn the coveted Astronaut's wings. Upon reentry "he pulled the sky dart's nose 23 degrees higher than its tail, falling spread-eagle style to slow descent, a maneuver future space ships may employ" according to the Associated Press coverage of the story. (Note: The Space Shuttle reenters with its nose up at an angle of 40 degrees.)

B.R. Schuster, McDonnell Aircraft engineering draftsman, paints Faith 7 on Spacecraft #20.

Gordon Cooper discusses a point with Mercury Flight Operations Chief Christopher Columbus Kraft during a preflight Mission review.

President John F. Kennedy pins the NASA Distinguished Service Medal on Cooper at the White House as Vice President Johnson and the other Astronauts watch.

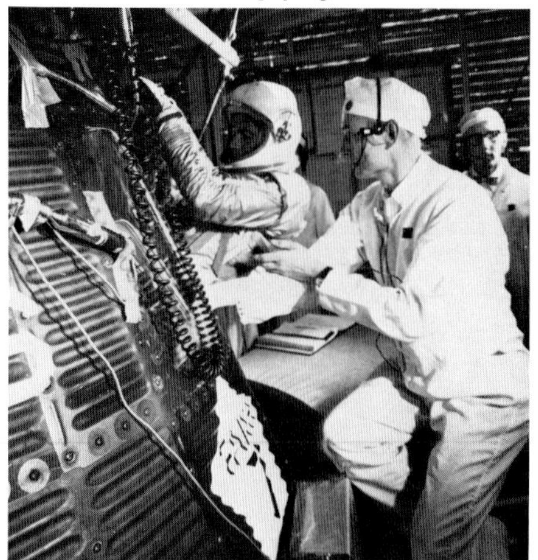

Gordo is helped into Faith 7.

HOME EDITION

The South Bend Tribune.

LAUNCH DELAYED ONE DAY

U. S. Lines Get Plane Fare Help

Tracking-Radar Trouble Blamed

U.S. Troops Ready If Violence Returns To Birmingham

BRITISH TRY TO ENFORCE HIGHER RATE

NEGROES RIOT IN NASHVILLE

HOME EDITION

The South Bend Tribune.

COOPER SEEKS U.S. RECORD

Capsule Whirls Free on 22-Orbit Journey

BRITISH HOLD PHYSICIST AS RUSSIAN SPY

Alabamans Want Troops Withdrawn

Space Ride Quite a Thrill, Astronaut Tells Tracker

"Gordo", as his friends called him, waited for launch on top of his venting booster for over 6 hours before a dirty fuel line in the gantry diesel and a bad C-Band radar system in Bermuda forced a cancellation. "Are you sure you can't run out and light the fuse with your little Zippo?" he asked Paul McDonald in Mercury Control. "Naw, I guess not today, Gordo," was the reply. As he crawled out of Faith 7 he said, "I was just getting to the real fun part...it was a very real simulation."

Cooper's official NASA Mission was "to determine the effects of extended space flights on man." Many scientists and doctors felt that the pooling of blood in a person's extremities would preclude long flights. This uncertainty was added to by the sickness that the Russian cosmonauts had experienced on their long-duration flights. On Orbit 3 he ejected a small 6 inch satellite with xenon strobe lights to test his ability to see a light of known intensity in space...important to future rendezvous missions in Gemini and Apollo.

FAITH 7

L. Gordon Cooper, Jr.
Shawnee, Oklahoma

Series: Mercury
Date: May 15, 1963
Crew & Age: L. Gordon Cooper, Jr. (36)
Distinction: Determined Effects of Extended Space Flight
Crew Insignia: NASA Patch
Spacecraft Name: "Faith 7"
Flight Duration: 22 orbits

"An awful lot of thought and symbolism had gone into all those earlier names. I felt a certain responsibility. I selected the name Faith 7 to show my faith in my fellow workers, my faith in all the hardware so carefully tested, my faith in myself and my faith in God. The more you study, the more you know all the scientific stuff, it correlates. It confirms religious faith.

"The name was painted on the side of the Mercury 7 spacecraft by the McDonnell Corporation at Cape Canaveral. The design was selected by me."

—Gordon Cooper

HOME EDITION
The South Bend Tribune.

COOPER GOING ALL THE WAY

Textbook Flight Uses Little Fuel, Oxygen

BUSINESSMEN BACK PACT IN ALABAMA

Much of the credit for the 22 orbit success of Faith 7 must go to the men that NASA considered to be the "Mercury Spacecraft Inventors", Maxime A. Faget, Andre J. Meyer, Jr., William Bland, Alan B. Kehlet, Willard S. Blanchard, Robert G. Chilton, Jerome B. Hammack, Caldwell C. Johnson and Jack C. Heberlig. They were now hard at work on Gemini and Apollo. On his 14th orbit, high over the Pacific Ocean, Cooper voiced a prayer: "Father, thank you for the successes we have had in flying this flight. Thank you for the privilege of being able to be in this position... be in this wondrous place, seeing all of these many startling wondrous things you have created. Help guide and direct all of us that we may shape our lives to be good..."

HOME EDITION
The South Bend Tribune.

CAPITAL CHEERS COOPER

Turkey Crushes Attempted Coup

Astronaut Given Medal by JFK

ARMY AIDE FAILS TWICE IN 15 MONTHS

High Court Ruling May Spur Sit-ins

POPE SUFFERS

Farmers Vote on Wheat Curbs.

Drawling out a "thank y'all" in his Oklahoma accent, the 36-year-old astronaut accepted a medal and President Kennedy's thanks in a colorful and crowded ceremony in the White House Rose Garden. All the Astronauts except John Glenn, who was in Japan, were there. Kennedy said that Cooper "went furthest in space and did so on the anniversary of Lindbergh's flight to Paris." "I think before the end of the 60's", Kennedy went on, "we will see a man on the Moon, an American." James Webb, director of the space program made awards to G. Merritt Preston, Cape Canaveral Operations Chief for the flight; Christopher C. Kraft, Flight Director; Kenneth Kleinknecht, Program Director for Mercury Flights and Major General Leighton L. Davis, Defense Department support.

GEMINI

"...an official NASA christening named the program Gemini, after the Constellation that includes the twin stars Castor and Pollux. The name was first suggested by Alex Nagy of NASA Headquarters, and it caught on overnight."
—Gus Grissom

From "GEMINI" © Macmilliam Company

Nagy thought that "the Twins", as Gemini, one of the 12 constellations of the Zodiac was called, did a good job of symbolizing the two man crew, a rendezvous mission and its relation to Mercury. By an unlikely coincidence, since Nagy disclaims any knowledge of astrology, Gemini as a sign of the Zodiac is controlled by Mercury. Its spheres of influence include adaptability and mobility—two features that spacecraft designers had explicitly pursued—and, through its link with the third house of the Zodiac, all means of communication and transportation as well. Astrologically, at least, Gemini was a remarkably apt name, the more so since the United States is said to be very much under its influence."
—NASA

On the Shoulders of Titans
A History of Project Gemini

THE GEMINI PROGRAM

Between the pioneering flights in Mercury and the astounding successes of Apollo came Gemini. The Gemini Spacecraft was really little more than a two man version of Mercury with the support equipment moved "out back" into an instrument unit.

Gemini proved these unknowns that we had to accomplish before a moon Mission was possible:

1. Rendezvous and Docking.
2. Long Term Flights.
3. Multiple-hour work outside the spacecraft.
4. Pinpoint reentrys.
5. Advanced reliable systems (fuel cells, cryogenic storage of hydrogen and oxygen, ablative thrusters, onboard digital computers, inertial guidance systems and rendezvous radars).
6. Trained flight and ground crews.

Gemini was the time when America caught up and surpassed the Soviet Union in the Race to the Moon.

The Gemini cockpit was not large, as shown in this view of Gemini VI astronauts Wally Schirra and Tom Stafford.

The Mercury and Gemini capsules at the McDonnell-Douglas plant in St. Louis. As can be seen, Gemini was little more than a two-man Mercury capsule.

THE GEMINI SPACECRAFT

The author has owned three Volkswagens and has sat in the Molly Brown, the first Gemini flown in space. The two vehicles are nearly identical in dimensions, with Gemini lacking a back seat. The cabin section was a truncated cone 70 inches long and 90 inches in diameter. The adapter (equipment) section behind it was 7½ feet long and 10 feet in diameter at its back end. On top of the cabin section was a reentry and rendezvous section 5 feet long and a little over three feet in circumference. Gemini contained 1,230,000 parts and 10.5 miles of wiring. 5,961 McDonnell Aircraft people in St. Louis worked on its construction. The ejection seats were built by Weber Aircraft Division, Kidde, Inc.

THE GEMINI ROCKETS

A Titan II rocket boosted Gemini into space. They were built by the Martin Company. For rendezvous and docking practice and to boost the Gemini into higher Earth orbits an Agena rocket built by Lockheed Missiles & Space Company was used. And to lift the Agena into orbit an Atlas rocket built by General Dynamics Corporation filled the bill.

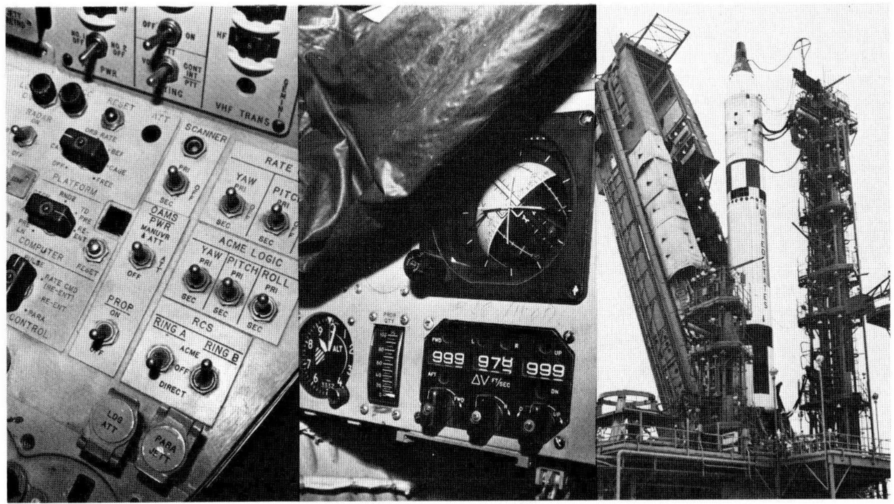

The console between Gus Grissom and John Young in the "Molly Brown."

"Gus Grissom's side of the "Molly Brown" cockpit. It was a relatively simple dashboard compared to the later Apollo displays."

The erector platform was lowered just 30 minutes before a Gemini was to be launched.

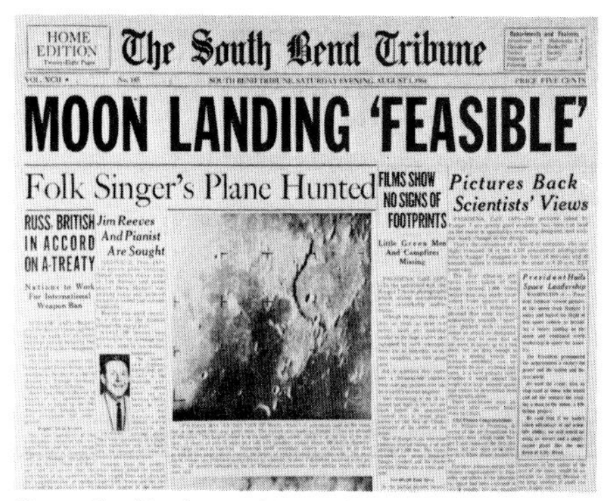

HOME EDITION

The South Bend Tribune

MOON LANDING 'FEASIBLE'

Folk Singer's Plane Hunted

Jim Reeves And Pianist Are Sought

RUSS, BRITISH IN ACCORD ON A-TREATY

FILMS SHOW NO SIGNS OF FOOTPRINTS

Little Green Men And Campfires Missing

Pictures Back Scientists' Views

Many exciting things happened in space between our Mercury and Gemini Programs. On July 31, 1964 the U.S. unmanned probe Ranger 7 successfully hit the moon near the crater Guericke in the Sea of Clouds, returning 4,316 closeup pictures termed "1000 times better than any made before." In future years both Apollo 12 and Apollo 14 would land nearby. The Associated Press coverage noted that the pictures showed objects as small as 3 feet across, but "there are no footprints, no extinct campfires, no little green men...the pictures are pretty good evidence that men can land on the moon in spaceships now being designed..." Russia launched its first 3 man mission aboard Voskhod 1 (Ruby) in October 1964 and Aleksey Arkhipovich Leonov performed the world's first extravehicular activity (spacewalk) aboard Voskhod 2 (Diamond) on March 18, 1965, just 5 days before our first Gemini flight, again stealing our thunder. The NASA Administrator had announced our launch date on February 5 in Lincoln, Nebraska.

HOME EDITION

The South Bend Tribune

KING LEADS 2ND DAY TREK

Weather Shift Raises Gemini Hopes

Launching on Schedule Given a Good Chance

HEAVY GUARD ALONG ROUTE OF MARCHERS

Find Sixth Bomb in Birmingham Search

Ranger 9 Streaking

HOME EDITION

The South Bend Tribune

ASTRONAUTS CHANGE ORBIT

'Historic' Maneuver Executed Over Texas

RUSS REPORT AID GOING TO N. VIET NAM

Astronauts' Sons Join Countdown

Shift in Course Done Manually

"You're on your way, Molly Brown," was the message from Cap Com Gordon Cooper when Gemini 3 rocketed off Pad 19 at 9:24 A.M. Tuesday morning, March 23, 1965. An hour and a half after launch Gemini 3 scored a "first" with an OAMS burn to circularize their orbit. Two more "burns" followed. Another "first" occurred when John Young asked Gus "You care for a corned beef sandwich, skipper?" John knew Gus didn't care for spacecraft food and so Wally Schirra had bought one he thought he'd like at "Wolfie's" on North Atlantic Avenue in Cocoa Beach and John had smuggled it aboard. NASA and Congress were not amused, but us ordinary folks thought it was hilarious.

The Soviets tried to steal Gemini's thunder with the first spacewalk just 5 days before the "Molly Brown" launch.

On Sunday, March 21, 1965 Gus Grissom and John Young watched the launch of Ranger 9 to the Alphonsus Crater on the moon. It was being sent to investigate the area as a future Apollo landing site and because of mysterious red flares spotted in the area which may have been volcanic activity. Ranger 8 had earlier landed in the Sea of Tranquility, future site of the First Moon Landing on Apollo 11. NASA watched the weather for the next day's launch as this subheadline details and Dr. Martin Luther King was leading the historic Selma to Montgomery Civil Rights March, in the freezing weather.

Chicago Tribune

FINAL

SPACE TRIUMPH FOR U. S.

Medicare Bill Is Approved

ASTRONAUTS NAVIGATE CAPSULE FOR 1st TIME

HOUSE GROUP ACTION HAILED BY PRESIDENT

Price Tag Placed at 6 Billions

Viet Gas Use Record: All 3 Tries Fail

Astronauts on Carrier After Flight

RANGER SENDS MOON TV INTO HOMES TODAY

Two-Man Gemini Crew Safe on Carrier

Hoosier Gus Grissom had become the first person to fly in space twice aboard Gemini 3, the space program originally called Mercury Mark II. Dr. Werner von Braun celebrated his 53rd birthday watching the launch and would go down in history as the man most responsible for putting America on the Moon. In tiny Mitchell, Indiana, Gus' home town, his Mom and Dad joined in a rain-soaked, but boisterous, parade celebrating the flight. A long line of brand new yellow school buses from the body works on the west side of Mitchell where Gus Grissom worked before going to Purdue University were in the parade plastered with signs, "Good Going, Gus!"

MOLLY BROWN

Virgil I. "Gus" Grissom
Mitchell, Indiana

John W. Young
San Francisco, California

Series: Gemini
Date: March 23, 1965
Crew & Age: Gus Grissom (38)
(Mercury—Liberty Bell)
John Young (34)
Distinction: 1st Man to Fly Twice Into Space (Gus Grissom)
Crew Insignia: NASA Patch
Spacecraft Name: "Molly Brown"

"In naming our Gemini 3 spacecraft, I always had in mind the unfortunate fate of my Liberty Bell 7, which sank like a stone when her hatch blew prematurely. I nearly went down along with her.

At first I kind of liked the idea of using an Indian name, say one of the tribes that once roamed Indiana, so I asked the research people at World Book Encyclopedia and Life Magazine to see what names those tribes had. They came up with the Wapashas, after whom the Wabash River is named. 'Great,' John and I agreed. We'd go into space aboard the Wapasha, a truly American name. Then some smart joker pointed out that surer than shooting our spacecraft would be dubbed the 'Wabash Cannon Ball."

Well, my Dad was working for the Baltimore and Ohio Railroad and I wasn't too sure just how he'd take to The Wabash Cannon Ball. How would he explain that one to his pals on the B & O?

At just about that time the Broadway musical comedy "The Unsinkable Molly Brown" was coming to its successful closing, and this gave me my clue. I'd been accused of being more than a little sensitive about the loss of my Liberty Bell 7, and it struck me that the best way to squelch this idea would be to kid it. So John and I agreed that we'd christen our baby 'Molly Brown.'

'Some of my bosses were amused;
some weren't.
'Come on, Gus, you can do better than that,'
the latter told me.
'What's your second choice for a name?'
'Well, I replied, 'what about the Titanic?'
'Nobody was amused, so Molly Brown
it was.'

—Gus Grissom
Command Pilot

From "GEMINI, a personal account of Man's venture into space." By Virgil "Gus" Grissom, Copyright 1968 By World Book Encyclopedia. The Macmillan Company, New York.

Gus Grissom studies his Flight Plan just prior to launch.

John Young (foreground) and Gus Grissom aboard "Molly Brown" just moments before ignition.

The first American to walk in space, Edward H. White II.

HOME EDITION

The South Bend Tribune

OK GIVEN FOR SPACE WALK

Maneuver Delayed From 2d to 3d Orbit

COLD, BLACK SPACE DEVOID OF DISTANCE

Families See Liftoff On TV Sets

Service Tower Disorder Mars Perfect Blastoff

At-a-Glance

Webb Bans Sandwiches From Spaceship Cabin

The launch from Pad 19 almost didn't come off when technicians couldn't get the booster erector to lower. It stuck at a 12-degree angle and it took them over an hour to find the problem in a junction box. Finally, at 10:16 a.m., Thursday, June 3, 1965 millions of people around America and for the first time in 12 European countries watched the launch via the Early Bird satellite. The new Manned Spacecraft Center in Houston was used for the first time for Gemini IV. As Gemini IV roared into space 8 year old Mike McDivitt looked up and observed: "Daddy's in there."

THE INDIANAPOLIS STAR

ASTRONAUT WALKS IN SPACE

'I'm Not Coming In,' Laughs White

'3½ Days To Go, Buddy,' Partner Chuckles Back

A Lovely Day To Hitch A Ride

Rendezvous Plan 'Scratched' For Using Up Fuel

Sight Russ Bombers In Red Viet Nam

Most people assumed that Ed White took his historic space walk because Leonov had done it 2½ months before, but it had been planned since 1962 and had been announced at the NASA press conference announcing the Gemini IV crew in July 1964. Another case of America giving the Russians a target date to beat in a space "first." Jim McDivitt attempted an "eyeball" orbital rendezvous with his booster, but NASA had not yet figured the secret of orbital mechanics—one had to slow down, drop into a lower orbit in order to speed up and catch a target object, a paradox, indeed. Ed White "swam" in space for 20 minutes over America, from the coast of California to south of Florida, at an altitude of 100 miles, and a speed of 17,500 miles an hour.

GEMINI IV

Edward H. White II
San Antonio, Texas

James A. McDivitt
Chicago, Illinois

Series: Gemini
Date: June 3-7, 1965
Crew & Age: James A. McDivitt (35)
Edward H. White II (34)
Distinction: First U.S. Space Walk
Crew Insignia: NASA Patch, American Flag
Spacecraft Name: "Little Eva" (Unofficial)

"Ed White and I used the American flag on our shoulders as our patch. This was the first time the American flag had been worn on a pressure suit and it has continued to be used there ever since. The original flags we had sewn on we purchased ourselves. Later on, of course, NASA made this an integral part of the pressure suit.

The name we selected for the vehicle was American Eagle. Unfortunately, we were unable to use this, and we were known as Gemini IV."

—Jim McDivitt
Command Pilot

The unofficial nickname of this spacecraft was "Little Eva" prompted, of course, by the extra-vehicular activity (EVA).

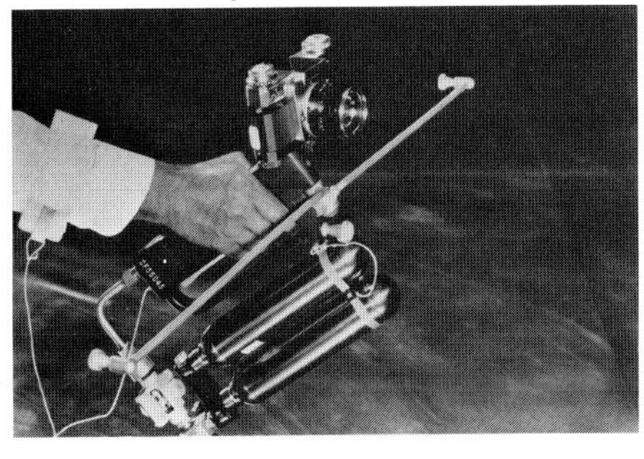

Ed White used a self-maneuvering unit powered by cold gas during his spacewalk. His camera was a stock Contarex 35-mm modified for spacesuit operation.

HOME EDITION

The South Bend Tribune

GEMINI NEARS U.S. RECORD

Astronauts Seek 34-Hour Endurance Mark

LBJ WANTS RUSS TO JOIN PEACE QUEST

President Calls For End of Aggression

Pilots Keep Sharp Eye On Ship's Fuel Gauge

At-a-Glance

White Seems Captivated As He Walks in Space

Viet Win to Be on Ground: Humphrey

Cardiovascular problems had cropped up with the final two Mercury missions, Gordon Cooper had staggered after recovery and some physiologists felt that man could not exist after prolonged spaceflights: "Don't you really know that these guys are going to stand up and pass out and might, indeed, 'die from this flight?" NASA people on hand at recovery were relieved "to see White do a jig-step, the air of tension immediately dissipated." The crew later inscribed a picture of themselves at recovery with "The day the straw men fell down."

HOME EDITION

The South Bend Tribune

SPACE TWINS' HEALTH A-OK

'Atom Session' Hears Branigin

Effects of Trip Surprisingly Few

2 SMASHER BILLS PUSHED BY GOVERNOR

Mission Control used three shifts for the first time with this much longer Gemini IV flight. Christopher Kraft acted as both Mission Director and Flight Director for the first shift, Eugene F. Kranz directed the second shift, and John Hodge the third. The flight lasted 97 hours and 56 minutes, proving that the Gemini spacecraft was "sea worthy."

NASA did not want to show the "8 Days or Bust" wording in case the flight had to be terminated early. Nevertheless, this did become the first Astronaut Mission Patch and in a NASA Administrator's memo of August 14, 1965 James E. Webb wrote:

1. "On GT-5 and future Gemini flights, such an identification (patch) may be worn on the right breast beneath the name plate of the Astronaut...this patch will be referred to by the generic name of the "Cooper patch."

2. "For GT-5, the "Cooper patch" will be the one submitted..."

3. "For Gemini flights after GT-5, the crew commander or senior pilot will be permitted to designate or design or recommend a "Cooper patch" for his flight, subject to approval by both the Director of the Manned Spacecraft Center and the Associate Administrator for Manned Spaceflight at NASA Headquarters..."

"Both Dr. Gilruth and I have a very real concern about the "8 Days or Bust" motto. I wish it could be omitted. If the Flight does not go 8 days, there are many who are going to say it was "busted." Further, whether we get the 8 days or not, the way the language will be translated in certain countries will not be to the benefit of the United States."

—James E. Webb
Administrator

GEMINI V

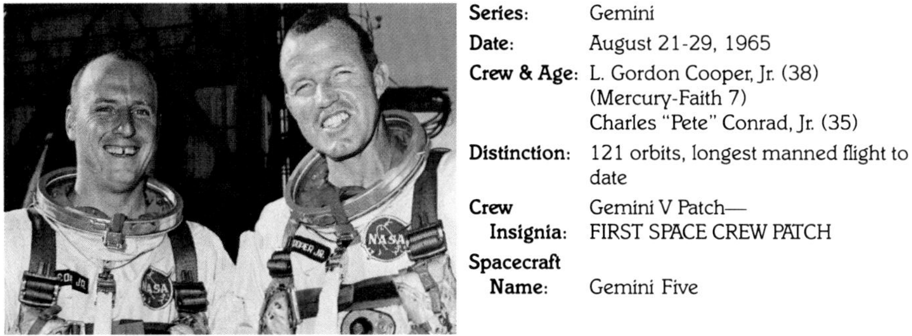

Charles Conrad, Jr.
Philadelphia, Pa.

L. Gordon Cooper
Shawnee, Oklahoma

Series: Gemini
Date: August 21-29, 1965
Crew & Age: L. Gordon Cooper, Jr. (38) (Mercury-Faith 7)
Charles "Pete" Conrad, Jr. (35)
Distinction: 121 orbits, longest manned flight to date
Crew Insignia: Gemini V Patch—FIRST SPACE CREW PATCH
Spacecraft Name: Gemini Five

The longest space flight prior to Gemini V had been the Soviet flight of Vostock V. Yasterb (Hawk), on June 14-19, 1963 by Valery Bykovsky. It had traveled 2,050,000 miles in its 81 orbits over a 119 hour period.

Cooper and Conrad's plea to NASA to let the Astronauts once again personalize their flights was finally approved. They reasoned that every flying outfit in the world has a patch and they should, too.

NASA approved their pioneer spirited design but made them cover up the "Eight Days or Bust" motto until after the flight in case they had to terminate it before that as the adjacent NASA correspondence shows.

Gemini V successfully broke all existing space records and covered 3,338,000 miles in its 119 orbits over a 190 hour period.

"Pete" Conrad wanted to name our Gemini spacecraft "Lady Bird" in honor of President Johnson's wife, but that was turned down. NASA was still not ready to let us name our spacecraft. Finally, Pete and I tackled them with the idea of a crew patch. My father-in-law had whittled a model of a Covered Wagon and I thought that was a good way to symbolize the pioneering nature of these early flights. The 8 days or bust wording just sort of followed along naturally. Pete and I had dinner with Mr. Webb and brought up the subject to him. We told him that NASA had taken our names away from us and we didn't particularly like it. We wanted to personalize our flight. And you know what, we won."

—Gordon Cooper
Command Pilot

THE CANTON REPOSITORY HOME EDITION

For 150 Years A Dependable Institution

CANTON, OHIO, SATURDAY, AUGUST 21, 1965

COOPER, CONRAD BLAST INTO SKY

Sheriff's Aide Held

Student Slain, Priest Is Shot Down in Dixie

Titan 2 Starts U.S. Pair on 8-Day Jaunt

L. GORDON COOPER — CHARLES CONRAD

Job Corps Riot Injures 10

Police Guard Kentucky Center

Viet Troops Find Cong Still Firing

Red Guerillas Keep Popping Up During Mopping-Up Efforts

The longest Soviet flight to date had been Vostok 5's 119 hours and in total National Cumulative Person-Hours in space Russia led with 507 hours, 21 minutes. Gordon Cooper and Pete Conrad would break both of those records on Gemini V. They stayed in orbit for 8 days, long enough to fly to the moon, explore its surface and return—190 hours and 55 minutes. And America passed the USSR with a new National Record 641 hours and 23 minutes in space. Russia would not catch America in total Man-Hours in space for an astounding 13 years, aboard Soyuz 29 in June, 1978! Slow out of the starting blocks, Yankee ingenuity had once more paid off.

Cooper and Conrad received a red carpet welcome aboard the U.S.S. Lake Champlain recovery ship after splashdown.

GEMINI VII

James A. Lovell, Jr.
Cleveland, Ohio

Frank Borman
Gary, Indiana

Series: Gemini
Date: December 4-18, 1965
Crew & Age: James A. Lovell, Jr. (37)
Frank Borman (37)
Distinction: 14 days in orbit • First U.S. Space Rendezvous with Gemini VI
Crew Insignia: Gemini VII Patch
Spacecraft Name: Gemini Seven

"Gemini 7 was to be a two-week mission with mostly medical experiments being conducted. Therefore, we wanted an insignia that would signify medicine and endurance, much like a long-distance runner. We chose the torch as the emblem. If you notice, the torch seems to be in motion signifying the flight of Gemini 7 over a long period of time. The artwork on the Gemini 7 patch was done by NASA Artists."

—Jim Lovell
Pilot

Gemini VII covered 5 million miles on its 14 day journey, enough for 10 round trips to the Moon. It served as the target for the first American rendezvous as Gemini VI came to within 1 foot of it and conducted sustained co-orbit. It was launched before Gemini VI in a change of plans due to the failure of the original Gemini VI Agenda target vehicle.

"The Spirit of '76" was what we ended-up calling Gemini's VI and VII. Because its Agena target vehicle blew up during launch from Pad 14, Gemini VI's launch from Pad 19 had to be scrubbed. Frank Borman and Jim Lovell, scheduled for Gemini VII, watched the abort and then talked to two McDonnell people, Walter Burke and John Yardley. Burke asked, "Why couldn't we launch a Gemini as a target instead of an Agena?" And so it was that President Johnson gave White House approval for the attempt and on December 4, 1965 Gemini VII blasted off as Jim Lovell shouted to his cabin-mate Borman, "We're on our way, Frank!"

A grizzled pair of space veterans after 14 days together in an area the size of the front seat of a Volkswagen.

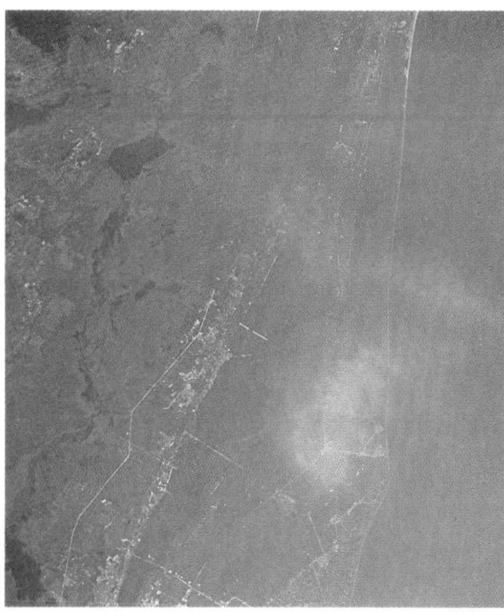

Borman and Lovell snapped a photo of the Kennedy Space Center as they flew overhead.

In feminique today—
HOW TO MAKE YOUR OWN CHRISTMAS ORNAMENTS

Chicago Tribune

CITY FINAL

GEMINI 6 LAUNCHING FAILS

Booster Ignites, but False Signal Cuts Off Power

Charge Murder in Bar Fire

SET FLAMES WHEN ANGRY, PATRON SAYS

13 Die, 22 Hurt in Disaster

NATURE NOTES

MARINES PUSH UPHILL FIGHT AGAINST REDS

Jets Blast Cong Caves, Tunnels

15 ON CARRIER

Rabu Hits U. S.

THE WEATHER

Then on Sunday, December 12 Astronauts Wally Schirra and Tom Stafford got a very unpleasant surprise when they attempted to roar-up into orbit to chase Gemini VII. At 1.2 seconds after ignition "an electrical tail plug dropped from the base of the booster rocket and the malfunction detections system shut off the engines." Wally Schirra decided not to pull the "D-ring" for ejection and thus did not disturb the integrity of the Gemini. Had the rocket risen even a few centimeters the shutdown would have brought 150 tons of volatile propellants encased in a fragile metal shell crashing back to Earth in a holocaust from which the crew could not have escaped. Wally Schirra kept his cool and did not pull the "chicken switch."

GEMINI VI

Thomas P. Stafford
Weatherford, Oklahoma

Walter M. Schirra, Jr.
Hackensack, New Jersey

Series: Gemini
Date: December 15-16, 1965
Crew & Age: Walter M. Schirra, Jr. (42) (Mercury-Sigma 7)
Thomas P. Stafford (35)
Distinction: First U.S. Space Rendezvous (with Gemini VII)
Crew Insignia: Gemini VI Patch
Spacecraft Name: Gemini Six

"I designed the patch to locate in the 6th hour of celestial right ascension. This was the predicted celestial area where the rendezvous should occur (in the Constellation Orion). It finally did occur there. The flight patch had an Agena target rather than a Gemini as in real life. Notice the "6" connecting the stars around Orion as well as the catchy name "Gemini 6". No names were permitted after "Molly Brown" (the unsinkable) until Spider in Apollo 9." —Wally Schirra
Command Pilot

"We were up there aiming for the rendezvous and when we first saw our rendezvous vehicle, Gemini VII, glittering in the reflected light of the sunset, it was right between Sirius and the twins, just exactly where we had placed it on the patch." —Tom Stafford
Life Magazine

Gemini VI was launched 11 days after Gemini VII due to three unscheduled delays. Their original Agena target failed to go into orbit, scrubbing their lift-off on October 25, 1965. NASA then decided to have Gemini VI rendezvous with Gemini VII. This was accomplished on December 15, 1965. Gemini VI was launched second because it had the radar. Gemini VII was 1,400 miles ahead of Gemini VI and 188 miles high and it took Gemini VI 6 hours (4 orbits) to catch up. They rendezvoused over the Western Pacific at night and stayed together for 7 hours (5 orbits).

GTA-6 on the patch refers to the original flight configuration of Gemini-Titan-Agena. On such a GTA mission the Titan rocket boosted the Gemini spacecraft into earth orbit where they separated. The Gemini then took off in pursuit of the Agena target rocket that had been launched earlier. Once it found the Agena the Gemini would attach itself to one end; this is called docking. Although the Agena was not used on this Gemini 6 flight the same procedure had been planned and it was too late to change the crew's patch design.

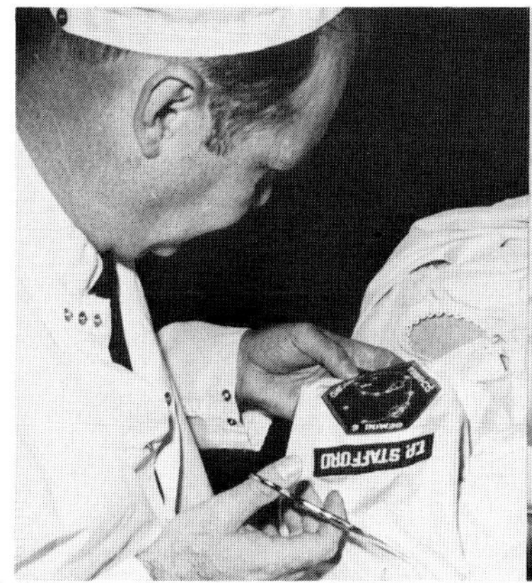

The Gemini VI patch is sewn on Tom Stafford's spacesuit.

Training for a flight was not all fun as this photo of Wally Schirra shows during emergency egress training.

The first successful true rendezvous in space.

Jim Lovell at breakfast the morning of his first launch. He would go on to become the most traveled man in history during 4 spaceflights.

Schirra and Stafford got in some good-natured kidding. They and Jim Lovell sitting next to Borman were all U.S. Naval Academy graduates, Borman was a West Point man. (Note sign in window.)

THE CANTON REPOSITORY HOME EDITION

Red China Nuclear Stockpile Seen in 2 Years

GEMINI 6 STARTS CHASE IN SPACE

NATO Hears Of Increasing Peking Might

Sky Is Stage For Historic Stalking Try

Councilmen Scan

Chicago Tribune

HOME

4 HOLD SPACE RENDEZVOUS

SHIPS ONLY 6 FEET APART

GO AROUND EARTH IN FORMATION; GEMINI 7 CREW SEEN BY VISITORS

SPACE WIVES LISTEN IN ON RENDEZVOUS

THE SALESMEN

Find Downed Airplane, 2d Is Missing

BOMB HANOI'S BIG SOURCE OF ELECTRICITY

Jets Hit Plant Near Haiphong

2 ECONOMISTS DIFFER ON HIKE

U.I. Trustees Expand Studies of Religion

Finally, at 8:37 a.m. on December 15, 1965 Wally Schirra urged his spaceship, "For the third time, Go!" And go it did. Up above in Gemini VII Borman and Lovell finally saw VI's contrail as they passed over Tananarive and got a brief glimpse of the chase spacecraft. They put on their spacesuits and waited for company to arrive.

At an altitude of 186 miles and in the terminal stages of the rendezvous the Gemini VI-A astronauts saw the stars Castor and Pollux in the Gemini Constellation aligned with their sister ship, Gemini VII, exactly where the target vehicle had been located in their Mission Patch designed months before. Schirra and Stafford had answered another important piece of the Lunar Mission puzzle.

Borman and Lovell at the launch pad wearing the new lightweight spacesuits for their long duration flight.

The happy astronauts congratulate one another following their pioneering rendezvous. Aboard the U.S.S. Wasp.

New Steel, Oil Output Records Seen . . . Column 2

THE HAMMOND TIMES

Home Newspaper of the Calumet Region

Gemini 6 Down; Crew Safe

Mission Hailed Abroad

Soviet Union Comment Slow

Carrier Overshot 12 Miles

Astronauts Both Fine

Its Mission over, Gemini VI came back down to Earth after just 25 hours and 51 minutes in space. But not before Tom Stafford excitedly reported: "Gemini VII, this is Gemini VI. We have an object, looks like a satellite going from north to south, probably in polar orbit...looks like he might be going to reenter soon. Stand by one... you just might let me try to pick up that thing." The strains of "Jingle Bells" came over the radio, with Schirra on a small 4 hole harmonica and Stafford jingling small bells. The Spirit of Christmas glowed.

THE CANTON REPOSITORY HOME EDITION

For 150 Years A Dependable Institution

SPACE CHAMPS HOME SAFELY!

Ordinance Held 'Unreasonable'

Rezoning of 25th st NW Site For Apartments Is Ordered

Team Weary But Jubilant Over Record

For the Newspaperboys

U.S. Skeptical of Offer

U.S. Asks Hanoi For Clarification Of Peace Feeler

With the excitement of rendezvous over, life returned to normal for several more days aboard Gemini VII. Frank Borman read some of Mark Twain's "Roughing It" and Jim Lovell read "Drums Along the Mohawk". Finally, after 14 days in space and with the popular "Going Back To Houston" ringing in their ears the duo returned to Earth. America had orbited 10 men since March.

The Atlas/Agena target vehicle roars aloft at 10 a.m.

The Gemini VIII launch at 10:40 a.m. with Neil Armstrong and Dave Scott aboard.

HOME EDITION

The South Bend Tribune

MOON CRAFT PASSES TEST

Marines Repel Viet Cong Attack

APOLLO SHIP GOES ALOFT, LANDS IN SEA

Reds Continue Pressure Tactic

GHANA'S NEW LEADERS FREE 450 IN JAILS

BULLETIN

Five Killed In Highway Accidents

While those in the Gemini Program were pressing on, solving and proving the intricacies of space travel, the people in the Apollo Program were readying the hardware for the day men would fly it. On February 26, 1966, just a few weeks before the Gemini VIII mission the first test of the Apollo Saturn IB rocket was successfully conducted at Cape Kennedy, as it was then being called. An unmanned Apollo flew a 5,300 mile test course and was recovered in the South Atlantic near Ascension Island.

THE CANTON REPOSITORY

HOME EDITION

For 150 Years A Dependable Institution

Minerva TRW Gets Alliance Bonney Plant

GEMINI 8 TARGET ROARS ALOFT

Production Start Seen In 2 Months

2 Astronauts Ready for 'Go' On Own Shot

Two Killed in New Los Angeles Riot

Massive Police Force Halts Rampage After 6 Hours

Gemini VIII would try again at what Gemini VI had failed to do, rendezvous with a target Agena rocket. The Agena was launched from Pad 14 at 10 a.m., 40 minutes later Gemini VIII with Neil Armstrong and Dave Scott aboard would thunder off. Scott was to do a spacewalk 20 hours later, to the back of the Gemini adapter section, using a special backpack and maneuvering gun. Fate would intervene, however.

GEMINI VIII

David R. Scott
San Antonio, Texas

Neil A. Armstrong
Wapakoneta, Ohio

Series: Gemini
Date: March 16-17, 1966
Crew & Age: Neil A. Armstrong (35)
David R. Scott (33)
Distinction: 1st docking of 2 orbiting spacecraft (Agena 8 target)
Crew Insignia: Gemini VIII Patch
Spacecraft Name: Gemini Eight

"The light from the twin Gemini stars is split by a prism in a spectrum of color spelling "Gemini VIII." This indicates that the flight objectives cover the complete sprectrum of Gemini program objectives (i.e. rendezvous, docking, EVA and experiments).

—Neil Armstrong
Command Pilot

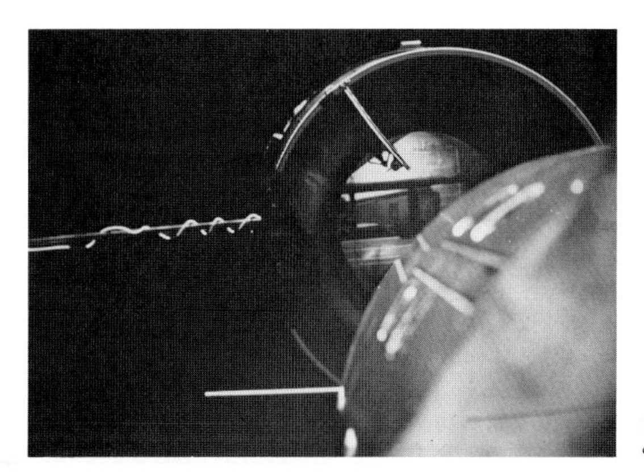

The first docking in space before the trouble started on Gemini VIII.

Chicago Tribune

THURSDAY, MARCH 17, 1966

TROUBLE IN GEMINI FLIGHT

Space Ship Docks and Then Backs Off

McCormick Pl. Is a Symbol of City Attractions

COPS PATROL WATTS AFTER NEGROES RIOT

2 Slain, 26 Hurt in Outbreak

'TIS AN IRISH WORLD TODAY—

USES UP 1 THRUSTER NEEDED FOR LANDING

We'll Be 1st with Man on Moon, Johnson Says

Order Craft to Make Landing at Once

The first part of the mission went off as advertised. Scott and Armstrong began their chase of the Agena exactly on time from Pad 19. And while they maneuvered to close the distance, Neil had time for some chicken and gravy casserole, fruit juice and brownies. At 5 hours and 14 minutes they rendezvoused with the Agena. Then at 6 hours and 32 minutes they got the go-ahead and docked successfully, another U.S. first. Shortly thereafter, things began to go sour. The two docked vehicles suddenly started to roll out-of-control. Fighting for control, the two quickly hit the undocking button. But the trouble persisted.

Chicago Tribune

FRIDAY, MARCH 18, 1966

SPACE HEROISM ON TAPES!

How 2 Tamed Gemini on Wild Ride

End Over End for 30 Minutes in Bucking Craft

STUDENTS ASK RED OUSTERS IN INDONESIA

ANOTHER ATTEMPTED HOOK-UP IN SPACE

HUNT N. VIET REGIMENT IN D ZONE DRIVE

Show Astronauts Did Right Thing at Right Time

Navy Destroyer Takes Crew to Okinawa Base

Thousands Riot in Jakarta

CITY PAT'S DAY OUTDOES IRISH

Chicago Gives Birth to New Concepts in Religious Practices

CLAY APPEAL TURNED DOWN

After backing away from the Agena, the spacecraft began to whirl at a dizzying rate of one revolution a second. Neil and Dave were having trouble seeing the panel dials and their physiological limits seemed near. They were dizzy and their vision was blurred. Reacting quickly, they were forced to use their reentry control system to bring Gemini VIII under control. Shortly thereafter they made an emergency landing in the Pacific east of Okinawa and south of Japan. They were rescued by the U.S.S. Destroyer Leonard F. Mason. After a month of testing, engineers at McDonnell's in St. Louis traced the problem to an electrical short on thruster #8. The flight had lasted 10 hours and 41 minutes. It had been a planned 3 day mission.

The "Angry Alligator"—the fiberglass cover failed to open, thwarting the docking exercise.

HOME EDITION

The South Bend Tribune

ASTRONAUTS DIE IN CRASH

U.S. Marines Battle Near Red Border

JET HITS ROOF, KILLING CREW FOR GEMINI 9

Charles A. Bassett, Elliot M. See Perish

Battle Zone 55 Miles From Communist North

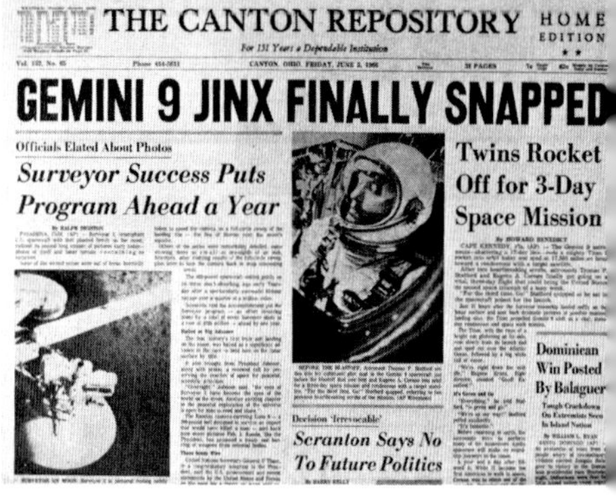

THE CANTON REPOSITORY

HOME EDITION

GEMINI 9 JINX FINALLY SNAPPED

Officials Elated About Photos

Surveyor Success Puts Program Ahead a Year

Twins Rocket Off for 3-Day Space Mission

Dominican Win Posted By Balaguer

Decision 'Irrevocable'

Scranton Says No To Future Politics

In October 1965 Elliot See and Charles A. Bassett II learned that they would fly Gemini IX, and that their back-ups would be Thomas Stafford and Eugene Cernan. One bright winter morning the 4 astronauts climbed into a pair of T-38's in Houston for the flight to St. Louis where their Gemini IX spacecraft was being built. See and Bassett in one, Stafford and Cernan in the other. Missing most of the runway due to overcast in St. Louis, See circled to the left to make another pass. Stafford climbed up to make another instrument approach. Elliot See had chosen to try to keep the field in view. Suddenly, the T-38 plunged into the very building in which their Gemini IX spacecraft was being readied. Both See and Bassett were killed. "On Wednesday, March 2, 1966 Spacecraft No. 9, on its way to the flight dock for shipment to Cape Kennedy, passed an American flag flying at half-mast at the McDonnell plant. The next day Elliot See and Charles Bassett, attended by their fellow astronauts were buried in Arlington National Cemetery."

"For the third time, Go!" Tom Stafford repeated before Gemini finally was launched. On May 17 their Atlas-Agena target vehicle had plunged into the ocean instead of going into orbit. On June 1 the target rocket went into orbit, but a problem with the launch control equipment kept the Gemini from hitting its 40 second long launch window and the mission was scrubbed. Stafford later said: "Frank Borman and Jim Lovell may have more flight time, but nobody had more pad time in Gemini than I did!" By the time Gemini IX-A lifted-off, he had been in the two spacecraft (6 and 9) ready for launch a total of six times.

GEMINI IX-A

Thomas P. Stafford
Weatherford, Oklahoma

Eugene A. Cernan
Chicago, Illinois

Series: Gemini
Date: June 3-6, 1966
Crew & Age: Thomas P. Stafford (35) (Gemini VI)
Eugene A. Cernan (32)
Distinction: 3 Rendezvous • Spacewalk • Most Accurate Splashdown
Crew Insignia: Gemini IX-A Patch
Spacecraft Name: Gemini Nine

"Gemini IX was an Agena rendezvous flight with an extended spacewalk. The large Roman numeral IX serves as the background for the "block design" Gemini-Agena typifying the docking profile. The tether, with an Astronaut "floating" on the end, forms the Arabic numeral 9 (as an ingredient of subtlety). The colors and shield design were to give it a thread of uniqueness. The final design was chosen over dinner with Stafford's and Cernan's wives being somewhat influential. The Agena, as history notes, became the "Angry Alligator"—what a patch that would have made."

—Gene Cernan
Pilot

This was the first flight in which a backup crew had been used in America's manned space flight program. It is for that reason that the flight carried the suffix "A".

Astronauts Elliot M. See, left, and Charles A. Bassett II, the original crew for Gemini IX. They died during a crash into the very building their Gemini was being built in.

THE INDIANAPOLIS STAR

CERNAN MAY UNHOOK SHIELD

Two Immolations Renew Viet Crisis

Buddhists Drive To Topple Ky

Successful Blastoff . . . Docking Problem

Officials Ponder Peril In Today's Walk In Space

Astronaut Sends Wire To Seniors

"The Angry Alligator" was the way Tom Stafford described the rendezvous target. The launch shroud covering the docking port had only partially opened. There would be no docking on Gemini IX. For awhile there was talk of Gene Cernan walking in space to try to free the shroud, but this was considered too dangerous. Finally, the duo began their equiperiod rendezvous.

Amberoid Upset Winner in Belmont

Details in Sports Section

Sox Beat Senators, 6-0; Cubs Defeat Reds, 5-3

Chicago Tribune

CERNAN SET FOR SKY WALK

Hood in a Panic: Giancana's Bid a Desperate One

MOB IN S. VIET TRIES TO BURN YANK'S OFFICE

Play It Cool on Year's Hottest Day

TEAM RESTS FOR COMPLEX TASKS TODAY

Recorder, Fuel Gauges Balk

McClory Urges House Subpena Justice Aids

Caged Mobster Was Clawing for Way Out

Buddhists Blast Johnson Policy

About 45 hours into the Mission Gene Cernan opened his hatch and stood in the opening, looking out at infinity. His spacewalk lasted 128 minutes and was marred by the fogging of his faceplate, but he did accomplish most of his objectives. This and the failure to dock marred an otherwise excellent mission highlighted with the three rendezvous exercises. On June 6 as their spacecraft rolled in the gentle swells of the Atlantic Ocean, Stafford and Cernan raised their arms and thumbed a ride on the approaching Navy ship U.S.S. Wasp. Splashdown occurred just 769 yards from the Wasp, America's most accurate landing up to that time.

A multiple exposure shows how the Gemini erector was lowered just before launch.

The patch includes two orbiting vehicles, the Gemini spacecraft and the Agena target vehicle. Twin Gemini stars, Castor and Pollux, represent the program and the crew. The Agena X was the first target on which the American flag was painted. Rendezvous and docking were important objectives of this flight. The crew located and docked with their Agena X target and later with Neil Armstrong and Dave Scott's Agena VIII. Mike space-walked over to Neil and Dave's Agena and retrieved a Meteorite package. This was the Agena that Armstrong and Scott had been docked with when they tumbled out-of-control.

Police Pledge Renewed Fight To Save Longevity Pay Plan – Page 1B

The News-Sentinel

CITY EDITION

SPACE LAUNCH COUNTDOWN BEGINS

Gemini Crew Awaits Liftoff

Wilson Fails To Find New Peace Hopes

U.S. Planes Pound Vital Red Oil Depot

Nurse Views Speck Under Heavy Guard

Big Marine Drive Dies Near Border

U.S. Delivers Warning Note To Red Cross

Michael Collins wrote in his book "Carrying the Fire", what it was like waiting for lift-off as this headline recounts: "Our Atlas-Agena (target vehicle) is off now, we are told, and we breath a sigh of relief...we will need it...we cannot see it or any other part of the outside world, save a tiny patch of blue sky overhead. We are lying on our backs looking straight up, with the Atlantic just under our right elbows, our feet pointing north. Our trajectory will carry us straight up for awhile; then we will lazily arc over to the east (our right) and will achieve orbit one hundred miles above the Atlantic, lying on our right sides. Time to climb, five minutes forty-one seconds; distance traveled downrange, 530 miles; velocity at engine cutoff, 17,500 miles per hour."

HOME EDITION

The South Bend Tribune

DOCK, HITCH SPACE RIDE

Rioters Set Cleveland Fires

Negroes Battle Police, Firemen

Gemini Gets OK To Open Hatch

'Captured Yank Pilots Face Trial'

Klein Says Javits Helped Him

FILIPINOS BEG

Following a rendezvous and docking with their Agena target the Agena engine roared into life and carried the Gemini X crew to a new record altitude of 476 miles high breaking the former record of 307 miles held by two Russian Cosmonauts. John Young later said: "At first, the sensation I got was that there was a pop (in front of our eyes), then there was a big explosion and a clang. We were thrown forward in the seats. We had our shoulder harnesses fastened. Fire and sparks started coming out of the back end of that rascal. The light was something fierce, and the acceleration was pretty good...the shutdown was just unbelievable. it was a quick jolt...and the tailoff...I never saw anything like that before...sparks and fire and smoke and lights."

GEMINI X

John W. Young
San Francisco, California

Michael Collins
Rome, Italy

Series: Gemini
Date: July 18-21, 1966
Crew & Age: John W. Young (35) (Gemini III)
Michael Collins (35)
Distinction: New World Altitude Record-476 Miles High • 1st Dual Rendezvous (Agena 8 and 10 targets) • 2 EVA's by Mike Collins • Used target as propulsion unit following docking
Crew Insignia: Gemini X Patch
Spacecraft Name: Gemini Ten

"On Gemini 10, which in my view has the best looking insignia of the Gemini series, artistic Barbara Young had developed one of John's ideas and come up with a graceful design, an aerodynamic X devoid of names and machines."

—Mike Collins
Pilot

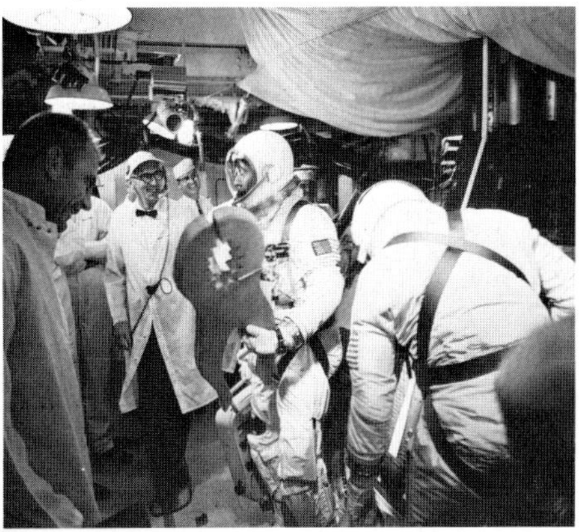

The crew was presented an over-sized pair of pliers by the White Room crew for any unforeseen emergencies. They were not taken aloft, but the humor broke the pre-launch tension.

HOME EDITION

The South Bend Tribune

TROOPS PATROL CLEVELAND

Collins Prepares for Stroll

Second Killed In Riotous Night

POPE PAUL ADDS VOICE TO APPEALS

Spacemen Chase Target Satellite

Things were not as tranquil on Earth as they were in space, as this front page points out. Nevertheless, Mike Collins walked in space from Gemini X over to the Agena from Gemini VIII that they had rendezvoused with after undocking from their own Agena. He took the S-10 experiment package off of Neil Armstrong and Dave Scott's Agena and crawled back into his spacecraft tangled up in his umbilical. John Young remarked that it made "the snakehouse at the zoo look like a Sunday school picnic." The EVA lasted 38 minutes. Mike also did a stand-up EVA of 49 minutes on the mission. Young and Collins landed in the western Atlantic after 70 hours and 46 minutes of flight.

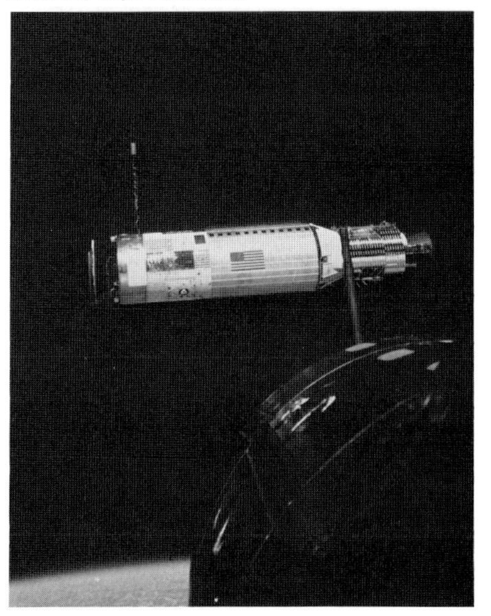

Man's highest flight at the time. 850 miles over western Australia, looking west from Perth to Port Darwin.

'Control' Protests By Doctors Against Medicare Flares Again—Page 15A

The News-Sentinel

CITY EDITION

Gemini Astronauts Complete Man's Fastest Space Linkup

Russ Claim Ex-priest Gets Asylum

Ky Says Vote Spells End of Reds' Threat

Biggest Air Assault Hits Top Areas

HOME EDITION

The South Bend Tribune

GEMINI SOARS TO 850 MILES

Rolvaag Whips Party in Minnesota

GOP's Joe Martin Loses Bid for 'One More Term'

UPSETS DFL CONVENTION NOMINEE 2-1

SPACECRAFT ALSO BREAKS SPEED MARK

We Ate Rats, Says POW

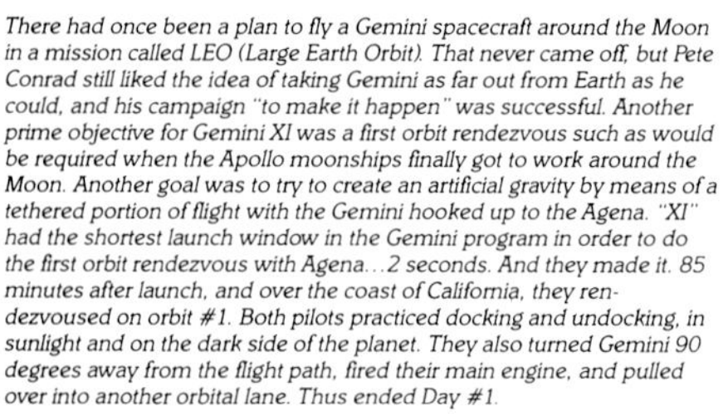

There had once been a plan to fly a Gemini spacecraft around the Moon in a mission called LEO (Large Earth Orbit). That never came off, but Pete Conrad still liked the idea of taking Gemini as far out from Earth as he could, and his campaign "to make it happen" was successful. Another prime objective for Gemini XI was a first orbit rendezvous such as would be required when the Apollo moonships finally got to work around the Moon. Another goal was to try to create an artificial gravity by means of a tethered portion of flight with the Gemini hooked up to the Agena. "XI" had the shortest launch window in the Gemini program in order to do the first orbit rendezvous with Agena...2 seconds. And they made it. 85 minutes after launch, and over the coast of California, they rendezvoused on orbit #1. Both pilots practiced docking and undocking, in sunlight and on the dark side of the planet. They also turned Gemini 90 degrees away from the flight path, fired their main engine, and pulled over into another orbital lane. Thus ended Day #1.

The next day Dick Gordon made a spacewalk to attach the 30 meter dacron tether (line) to the Agena. He sat astride the spacecraft nose, riding bareback, with his feet and legs wedged between the docked vehicles. "Ride 'em, cowboy!" Conrad shouted. On Day #3 he fired the Agena's main engine for 26 seconds. "Whoop-de-doo!" Conrad yelled, "that's the biggest thrill of my life." They watched the great round ball of Earth recede. What about orbital mechanics now? Were they going to stop, or continue out into deep space? "The world's round...you can't believe it," Conrad radioed... "I can see all the way from the end, around the top...about 150 degrees...everything looks blue!

They reached an altitude of 850 miles, and then made another burn to lower their orbit. Dick Gordon then made a two hour stand-up EVA in the hatch. It was so relaxing that both crewman catnapped with the hatch open. When they tried the tether experiment after undocking with Agena Conrad reported: "This tether's doing something I never thought it would

GEMINI XI

Richard F. Gordon, Jr.
Seattle, Washington

Charles Conrad, Jr.
Philadelphia, Pa.

Series: Gemini
Date: September 12-15, 1966
Crew & Age: Pete Conrad (36) Gemini V
Richard F. Gordon (36)
Distinction: Shortest launch window in Gemini—2 seconds • Achieved rendezvous with Agena on 1st orbit • 1st tethered flight—an experimental step toward building future space stations • New 850 mile altitude record • 2 EVA's (Gordon)
Crew Insignia: Gemini XI Patch
Spacecraft Name: Gemini Eleven

"Pete Conrad was able to convince NASA to let us take our Gemini spacecraft up to a new orbital record. Our Gemini XI patch depicts the various elements of our mission. The first orbit rendezvous (M equals 1) is marked by a star, so is our docking with the Agena target vehicle, our high orbit of 850 miles and my spacewalk. The Roman numerals "XI" are shown raising off the Earth as another indication of our new altitude record. We had the shortest launch window in the Gemini Program. (Richard R. Corley of the Gemini Program Office first suggested trying a first orbit rendezvous). Our window was just 2 seconds long and we only used a ½ second of it on launch. On reentry we let the computer fly us on in for the first time. All the previous flights had been flown in from about 72 miles altitude by the Commanders. This pinpoint accuracy was necessary for our future Moon Missions. The patch also shows the docked Gemini-Agena combination and my tethered spacewalk. Our Gemini-Agena tethered experiment produced an artificial gravity of two bodies orbiting together."

Both Pete and I are Navy men and used blue and gold for our patch since they are the Navy's colors."

—Richard F. Gordon, Jr.
Pilot

(Editor's Note: This practice proved very useful for Dick Gordon and Pete Conrad when they flew to the Moon together on Apollo 12).

"Ride 'em Cowboy"—Dick Gordon astride Gemini as he attached a tether 160 miles over the Atlantic Ocean on the 16th orbit.

do. It's like the Agena and I have a skip rope between us and it's rotating and making a big loop. Man! Have we got a weird phenomenon going on here. This will take somebody a little time to figure out." Strangely, although the spinning line was curved, it also had tension. Finally, the line straightened out. On September 15, 1966 after 71 hours and 17 minutes Gemini XI returned to Earth.

Buzz Aldrin at the Agena Work Station during his spacewalk.

THE WASHINGTON POST

Gemini-Agena Docking Accomplished

Radar Fault Complicates Maneuver

CAPE KENNEDY, Fla., Nov. 11—Astronauts James Lovell and Edwin (Buzz) Aldrin wrote a triumphant end to the Gemini program when they locked their orbiting spacecraft onto an unmanned Agena target vehicle over the Philippine Islands tonight.

Following their launch on Veteran's Day, Lovell and Aldrin discovered that their on-board computer was not working properly. And in one of the strangest coincidences of the Space Program, the one man that had worked on orbital rendezvous as part of his doctoral thesis at M.I.T. and who had developed a series of charts to be used by Gemini VI in case of just such a failure, was sitting in the right hand seat of Gemini XII. Buzz Aldrin had done both of these things and was even known to his friends as "Dr. Rendezvous." So using his own emergency charts and an eight-power sextant, he guided Jim Lovell to the orbiting Agena target.

Gemini in Race With Eclipse

Aldrin For Stand In Space

CAPE KENNEDY, Fla., Nov. 12 (AP)—Using a substitute flight plan, the Gemini 12 astronauts raced today toward a 13-second peek at a total solar eclipse and a 2½-hour "space stand" by pilot Edwin E. (Buzz) Aldrin Jr.

On the second day of their Mission, Lovell and Aldrin fine-tuned their orbit to take advantage of a solar eclipse across South America from north of Lima, Peru nearly to the southern tip of Brazil. The crew reported that they saw the eclipse "right on the money."

GEMINI XII

Edwin E. Aldrin, Jr.
Montclair, New Jersey

James A. Lovell, Jr.
Cleveland, Ohio

Series: Gemini
Date: November 11-15, 1966
Crew & Age: Jim Lovell (38) (Gemini VII)
Edwin E. Aldrin, Jr. (36)
Distinction: Rendezvous and 3 dockings • Longest EVA (Aldrin) 3 EVA's • Lovell most traveled man in history
Crew Insignia: Gemini XII—The Halloween Patch
Spacecraft Name: Gemini Twelve

"The Gemini XII patch was designed by Buzz and myself after trying to think of an appropriate patch for sometime. The flight was originally scheduled to fly on or about Halloween. That is why the patch had Halloween colors. The patch is also symbolic of the time of the mission with the spacecraft representing the hour hand pointing toward the Roman Numeral XII. Since this was the last flight prior to our Apollo moon missions, we included the familiar moon symbol on the left hand side of the patch. The artwork for this patch was done by McDonnell Douglas."

—Jim Lovell
Command Pilot

Jim Lovell presents a Million Dollar check to the White Room crew for their fine job preparing the spacecraft for launch.

Gemini Program Ends on Upbeat

ABOARD USS WASP, Nov. 16 (UPI) — Their troubled but record-setting Gemini 12 flight behind them, astronauts James Lovell and Edwin (Buzz) Aldrin today looked forward to their next trip in space — perhaps to the moon.

Their performance during the four-day flight — termed "magnificent" by NASA chiefs — closed out the series of two-man Gemini flights. But it opened the way for the Apollo moon-landing project.

Col. Adlrin, who spent a record 5½ hours outside the Gemini 12 hatch and space endurance champion Maj. Lovell planned to fly to Cape Kennedy later today.

There, where the final Gemini flight began Friday, they will undergo final briefings and medical checks before they return to their families in Houston.

Despite the rigors of the flight — and 22 minutes in their bobbing spacecraft before they were plucked yesterday from the whitecapped Atlantic, they bounded from their helicopter

The highlight of the Gemini XII Mission was the work done in space at the "busy boxes" by Buzz Aldrin. He torqued bolts, cut metal and used "Golden Slippers" to hold himself in place. Buzz also photographed Star Fields, retrieved a micrometeorite package and tied the Gemini and Agena together for another successful four hour tether experiment. At one point he even wiped off Jim Lovell's window, prompting the Commander to retort— "Hey, would you change the oil, too?" At 2:21 p.m. EST on November 15, 1966 Gemini XII splashed down close to the U.S.S. Wasp to end the highly successful Gemini Program. President Lyndon B. Johnson said that now "we know that America is in space to stay."

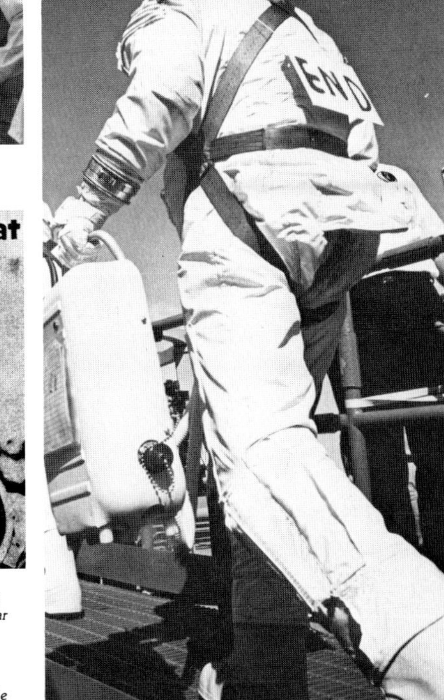

The crew of Gemini XII wore signs on their backs as they arrived at the Pad noting "The End" of the Gemini Program.

APOLLO

In July 1960 NASA was preparing to implement its long-range plan beyond Project Mercury and to introduce a manned circumlunar mission project—then un-named—at the NASA/Industry Program Plans Conference in Washington. Abe Silverstein, Director of Space Flight Development, proposed the name "Apollo" because it was the name of a god in ancient Greek mythology with attractive connotations and the precedent for naming manned spaceflight projects for mythological gods and heroes had been set with Mercury. Apollo was god of archery, prophecy, poetry and music, and most significantly he was god of the sun. In his horse-drawn chariot, Apollo pulled the sun in its course across the sky each day. NASA approved the name and publicly announced "Project Apollo" at the July 28-29 conference.

NASA SP 4402

The Apollo Program Emblem

The Earth and the Moon flank a large stylized letter "A" against a background of stars. The Constellation Orion, The Mighty Hunter, is positioned so that its three central stars, known as Orion's belt, form the bar of the letter "A". These stars are Mintaka, Alnilam, and Alnitak. The star shown above the Moon is Orion's shoulder, the red star Betelgeuse, and his other shoulder on the right top side of the "A" is the white star Bellatrix. Under the right side of the "A" is Orion's foot, the blue-white star Rigel and under the left side of the "A" is Orion's other foot, the blue-white star Saiph. Rigel is one of the Apollo Astronaut's 37 navigational stars. Between the limbs of the "A" is Orion's sword and in the center of the sword is the Orion Nebula, "a dense concentration of interstellar gases rich in newborn and young stars," according to Jastrow and Thompson—"Astronomy Fundamentals and Frontiers."

World Book Encyclopedia notes that "Orion was a mighty hunter in Greek mythology. He was the son of Poseidon (Neptune), who gave him the power to walk through the sea and on its surface.

The goddess Artemis (Diana) fell in love with the handsome Orion. Her brother, Apollo, did not like this, and plotted to destroy Orion. One day while Orion was swimming, Apollo walked by with Artemis. Apollo challenged her to hit the target bobbing in the water. Artemis, did not know that it was the head of her lover, and killed him with her arrow. Her sorrow was great and she placed Orion in the sky as a constellation."

The face on the Moon represents the mythical god, Apollo.

The following abbreviations are used on the Apollo flights to designate the primary job of the crew members: CDR = Commander; CMP = Command Module Pilot; LMP = Lunar Module Pilot. On lunar landing missions the CDR and LMP actually landed on the Moon while the CMP orbited overhead and "watched the store."

Apollo at night was an unforgettable sight. This is Apollo 8 the night before the first men flew to the Moon.

The Apollo Program

In his May 25, 1961 address to Congress, President John F. Kennedy said: "Now is the time...for this nation to take a clearly leading role in space achievement, which in many ways may hold the key to our future on Earth...this is not merely a race. We go into space because whatever mankind must undertake, free men must fully share."

James E. Webb, NASA Administrator, has said: "In the turbulent 1960's, Apollo flights proved that man can leave his earthly home with its friendly and protective atmosphere to travel out toward the stars and explore other parts of the solar system. Historians will find many lessons in the Apollo program for the managers of future large-scale enterprises. It was a new kind of national venture. Suddenly and dramatically it brought men of action and men of thought into intimate working relationships designed to solve a large number of extremely difficult scientific and technical problems. It was a major challenge to legislators, scientists, and engineers. The idea that if we can go to the Moon we can accomplish other feats long considered impossible has been firmly implanted in people's minds."

Planning for a lunar manned mission began in April 1957, and in July 1960 NASA named the program Project Apollo. Goals were:

1. To land American explorers on the Moon and return them safely to Earth.
2. To establish the technology required to meet other national interests in space.
3. To achieve for the United States preeminence in space.
4. To carry out a program of scientific exploration of the Moon.
5. To develop man's capability to work in the lunar environment.

The interior of the Apollo Command Module with about as much elbow room as the front seat of the family station wagon.

Apollo Command Module

This cone shaped cabin contained 15 miles of wiring and over 2,000,000 different parts. It was 12 feet high and 12 feet, 10 inches at its base. It weighed 12,392 pounds at launch and was constructed primarily of aluminum alloy, stainless steel and titanium. The honeycomb aluminum used in Apollo's inner crew compartment was 40% stronger and 40% lighter than ordinary aluminum. The Command Module used only 2000 watts of electricity, similar to the amount required by an oven in an electric range. Its crew panel displays included 24 instruments, 566 switches, 40 event indicators and 71 lights.

The Apollo Command and Service Module. The Astronauts rode in the shiny conical part on top. Note the large rocket exhaust bell.

"Snoopy" with legs folded is ready to be mated with the "stack" prior to flight. This was the Apollo 10 Lunar Module. Built by Grumman.

Apollo Service Module

This was the "extra room" that was attached to the backside of the Command Module. It contained the electrical power subsystem, fuel, the main propulsion engine and other necessary systems. On later Moon flights it also contained cameras and other scientific equipment used in lunar orbit. More than a mile-and-a-quarter of film would be used during the Apollo 15 Mission, for example. The Service Module was separated from the Command Module just prior to reentry into Earth's atmosphere. It was 22 feet, 7 inches long.

The Lunar Module

Looking somewhat like it was designed by a committee with no master plan, the LM (Lem) was designed to fly only in the vacuum of space. It was composed of two basic parts sitting atop spider-like shock-absorbing legs. The top portion was the crew cabin, the bottom the descent stage that also did double duty as a launch pad when the Astronauts decided to fire the ascent engine and leave the Moon's surface. The LM was 23 feet, 1 inch high and was also made of aluminum alloy.

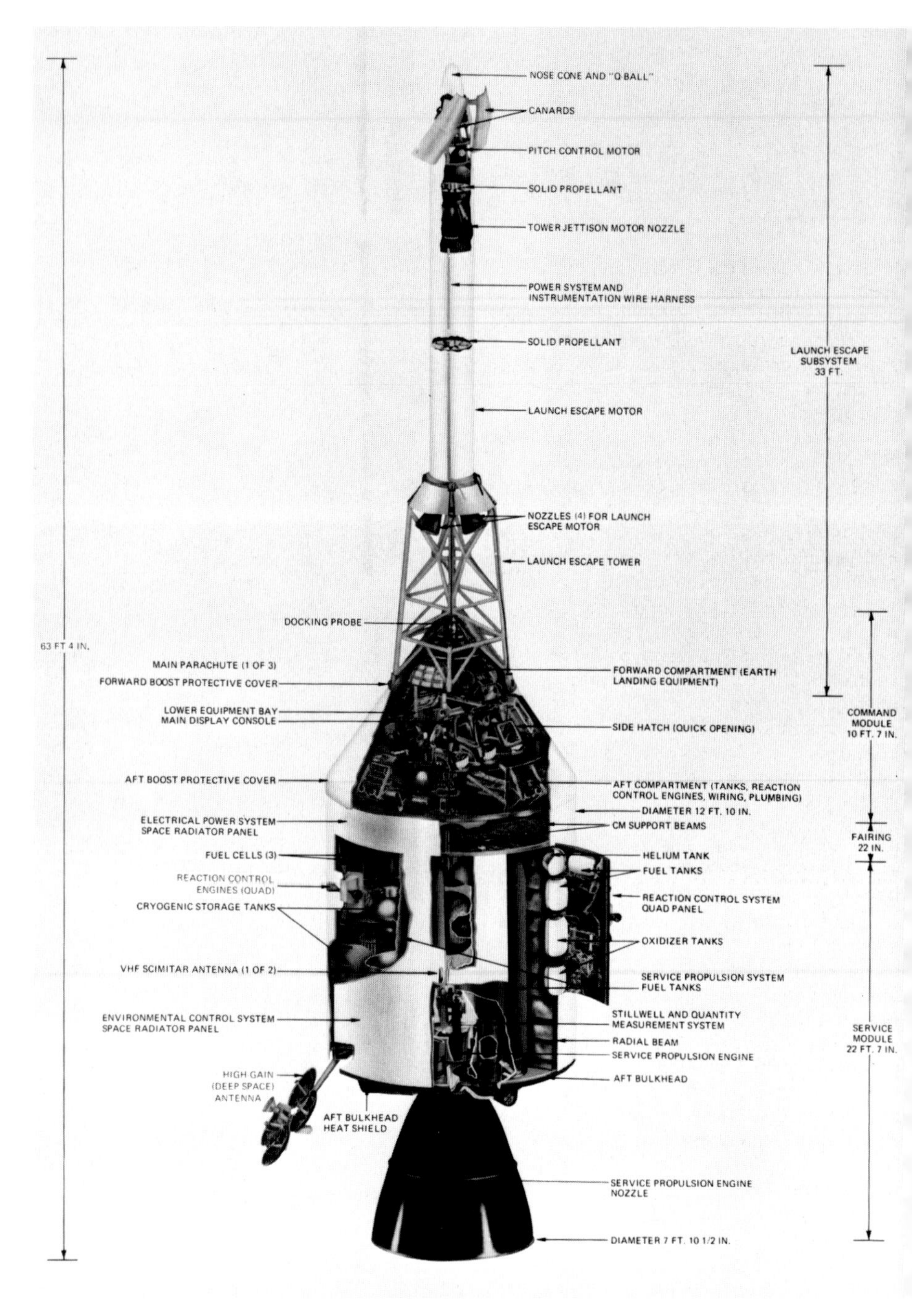

The Apollo Command and Service Module with escape tower.

The Saturn 5 Rocket

The goal was to place three Astronauts into orbit, plus all of the equipment and supplies they would need to fly to the Moon, land, explore, and return to Earth. To do this with a one stage rocket would mean that as it was running out of fuel its own dead weight would work against it. So a three stage rocket was designed by Dr. Werner von Braun and his men.

Stage #1 lifted the 6.4 million pounds off Earth and to an altitude of 38 miles in 2½ minutes. At that point the 363 foot tall rocket was moving 6,200 miles per hour and was 50 miles out over the Atlantic Ocean. It then shut off and the empty 300,000 pound 1st stage dropped off.

The 2nd stage then ignited and burned for 6½ minutes and carried the vehicle up to 108 miles and a speed of 15,500 miles per hour. It, too, dropped off and freed Saturn of another 95,000 pounds of dead weight. Saturn now had lost 95% of its original weight when the expended fuel was figured in.

Stage #3 ignited and burned for about 2 minutes up to an altitude of 115 miles and 17,500 miles per hour. Apollo was now in Earth orbit. The 3rd Stage did not drop off yet. After a trip or two around the World to check-out all systems, the 3rd stage was refired to boost the rocket toward the Moon at a speed of 25,000 miles per hour to escape Earth's gravitation.

The Command Module then separated, turned around, docked with the Lunar Module and pulled it out of its resting place in the top of the Saturn. The 3rd stage's work was then over and it went its own way out into deep space or to impact on the Moon. The Command and Service Module and the Lunar Module, now joined together at their heads, flew on to the Moon.

The Moon moves at a speed of 2,000 miles per hour in its orbit around the Earth, so in the time it took Apollo to travel out to the Moon, it had moved 165,000 miles. Apollo was aimed at where the Moon would be when it arrived, "leading it" much as a hunter does a flying meal of roast duck.

Lift-off of Apollo 10 shook the ground, rattled windows and left shock waves in its wake that the birdwatchers could feel against their faces three miles away. There were reports on some Apollo flights that the shock waves registered as far away from Cape Canaveral as New York City.

The Patch That Never Flew

SECTION A **THE JOURNAL-GAZETTE**

FORT WAYNE, INDIANA, SATURDAY, JANUARY 28, 1967

Grissom, White, Chaffee Die; Trapped In Fire On Apollo

Blaze Breaks Out During Rehearsal Of Feb. 21 Shot

U.S. Space Chiefs Vow To Go Ahead

Dillon Says 'Fair Bus' Bill Legal

Looting Breaks Out In Chicago;

Crews Push Storm Repair, Cleanup Task

HOME EDITION **The South Bend Tribune**

SOUTH BEND, INDIANA, SATURDAY EVENING, JANUARY 28, 1967 PRICE FIVE CENTS

ASTRONAUT "CRIED 'FIRE!'"

Storm Leaves Death Trail

ROOFS FALL; SHOVELERS' HEARTS FAIL

Heavy Equipment Of Guard Pressed Into Service

FAMILIES NEVER READY FOR DEATH

Indicates Brief Time for Action

'FAIR BUS' BILL RULED LEGAL

In the White House in Washington, D.C. diplomats from 60 nations were signing a new Peace-in-Space Treaty that President Lyndon Johnson described as "an inspiring moment in the history of the human race." Soviet Ambassador Anatoly F. Dobrynin told the East Room audience: "Let us hope we shall not wait long for the solution of earthy problems." The treaty aimed at preventing territorial or military rivalries in outer space and at blocking the orbiting of nuclear warheads. Astronauts Neil Armstrong, Jim Lovell, Dick Gordon, Gordon Cooper and Scott Carpenter were on hand along with rocket pioneer Dr. Wernher von Braun. The 2,000 word treaty was also aimed at preventing territorial claims in space, such as asserting national title to real estate on the Moon. It was quite appropriate that Neil Armstrong, who would later be our First Man on the Moon, was there. The happy atmosphere was shattered with the sudden news of the violent deaths of Grissom, White and Chaffee.

"If we die, we want people to accept it. We're in a risky business," Gus Grissom had said just weeks before the fatal fire aboard Apollo 1, "and we hope if anything happens to us, it will not delay the program. The conquest of space is worth the risk of life." On January 27, 1967 at Pad 34 Apollo 1 was going through an important "plugs out" test to see if the Moonship could sustain itself on its own. Grissom, White and Chaffee had climbed into the spacecraft about 1:00 p.m. and been sealed inside in a simulation of the actual countdown that was to occur the following month. Pure oxygen at a pressure of 16.7 pounds per square inch was pumped into the cabin. As they entered the spacecraft Gus had complained that it smelled funny, sort of like sour milk. The test went routinely during the afternoon with the usual communications glitches and the like. Then, just before sunset at 6:31 p.m. the technicians outside in the white room and elsewhere heard a cry over the Astronaut's radio circuit from inside the capsule: "There is a fire in here." It took 5 minutes to get the hatches open, by then the crew was dead, not burned to death as most people assumed—their bodies were protected by their spacesuits. They died from asphyxia due to inhalation of toxic gases. It was later decided that the fire started due to a short circuit near Gus' couch. There was no fire extinguisher in Apollo, a tragic oversight.

APOLLO 1

In Memoriam

Edward H. White II
San Antonio, Texas

Virgil I. "Gus" Grissom
Mitchell, Indiana

Roger B. Chaffee
Grand Rapids, Michigan

Series: Apollo

Date: January 27, 1967

Crew & Age: CDR. Gus Grissom (40) Mercury-Liberty Bell 7; Gemini 3-Molly Brown
CMP. Ed White (36) Gemini IV
LMP. Roger Chaffee (31)

The Apollo 1 crew died in their spacecraft atop a rocket sitting on their launch pad #34 three weeks before they were to have flown in space. They were conducting a plugs-out test.

Spacecraft Name: Apollo One

Crew Insignia: Apollo 1 Patch, U.S. Flag, NASA Insignia

"The Apollo 1 patch was designed by the crew as all others were and had been approved. Since this was destined to be the first Apollo orbital flight this was the prime theme. Names were not used after Mercury and the first Gemini flight except where radio call signs were required to differentiate between lunar modules and command modules. Therefore, Apollo 1 would never have been anything other than Apollo 1."

—Donald K. "Deke" Slayton
Director, Flight Crew Operations

100 PAGES **The South Bend Tribune**

APOLLO PROJECT STALLED

Work to Clear Area Highways

Disaster Causes Indefinite Delay

Assembly Buried By Flood of Bills

ONLY MAJOR ROADS ARE 'PASSABLE'

President Johnson led the nation in mourning for the fallen Astronauts. He sat with the Grissom family at the last rites on a frosty hillside in Arlington National Cemetery just a few hundred yards from the grave of President John F. Kennedy. Later in the day he also attended the burial of Roger Chaffee in Arlington. Ed White was buried in the cemetery of his beloved West Point Military Academy.

The Apollo 204 Review Board submitted its report on the fire on April 5, 1967. It concluded that the Apollo 1 atmosphere was lethal 24 seconds after the fire had started and that consciousness was lost between 15 and 30 seconds after the first suit failed. "Chances," the report stated, "of resuscitation decreased rapidly thereafter and were irrevocably lost in four minutes." The hatches were finally pried open five and a half minutes after the first alarm. They picked George M. Low to put the Apollo Program back together again. And at the Sea of Tranquility on the Moon the sun would rise only 33 more times before 1970. If we were to meet President Kennedy's National Goal much work had to be done. Roger, Gus and Ed's deaths had uncovered the fatal flaws in our Moonship...and 1,341 changes would be made in Apollo before it was deemed worthy of once more attempting to fly in space. 150,000 men and women would work for 21 months preparing Apollo. That they succeeded, and that the United States achieved its National Goal in July of 1969, is the Legacy of Apollo 1...the Legacy of Roger Chaffee, Gus Grissom and Ed White.

Cape Canaveral. . .where it all began. Apollo 7's spent stage high over its launch pad.

APOLLO 7

Donn F. Eisele
Columbus, Ohio

Walter M. Schirra, Jr.
Hackensack, New Jersey

Walter Cunningham
Creston, Iowa

Series: Apollo
Date: October 11-22, 1968
Crew & Age: CDR. Wally Schirra (45)
Mercury-Sigma 7; Gemini VI
CMP. Donn F. Eisele (38)
LMP. Walter Cunningham (36)
Distinction: First U.S. 3 man mission • 11 Days in Orbit, more man hours than all Soviet flights combined.
Crew Insignia: Apollo 7 Patch, U.S. Flag, NASA Insignia
Spacecraft Name: Apollo Seven

"The Apollo 7 design itself highlighted the earth orbital nature of the mission. It was our original intention to emphasize the *first* manned Apollo (Gus Grissom's flight) and the recovery from the fire on the pad aspects as well. We considered depicting a spacecraft rising from a ball of fire and calling it the Phoenix. The patch designed was subject to NASA approval and we abandoned the Phoenix theme feeling it would be rejected as in bad taste. I zeroed in on a circle (for the Earth) and an ellipse (for orbit). The orbital plane was tilted for artistic reasons." —Walt Cunningham
Lunar Module Pilot

A Mission Planning Meeting at the Cape three days before scheduled lift-off.

Breakfast on the morning of launch. America's pride rode with the Apollo 7 crew as they rose from the same launch pad the Apollo One crew had died on. Left to right: Kenneth Kleinknecht, Manager, Apollo Spacecraft Office, Astronaut Eisele, Astronaut Schirra, Deke Slayton, Director, Flight Crew Operations, Astronaut Bill Pogue and Astronaut Ron Evans.

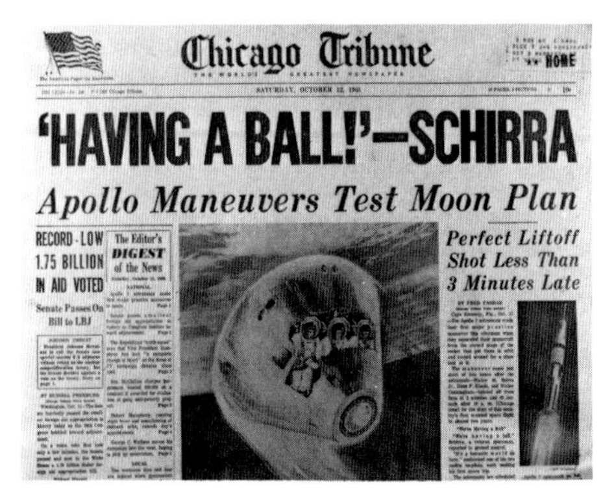

Chicago Tribune

HOME

SATURDAY, OCTOBER 12, 1968

'HAVING A BALL!'—SCHIRRA

Apollo Maneuvers Test Moon Plan

RECORD-LOW 1.75 BILLION IN AID VOTED

Senate Passes On Bill to LBJ

The Editor's DIGEST of the News

Perfect Liftoff Shot Less Than 3 Minutes Late

When Apollo 7 was launched a few minutes after 11:00 a.m. on October 11, 1968 it rose from the very same launch pad #34 that Gus Grissom, Ed White and Roger Chaffee had died on 21 months before. At age 45 Wally Schirra became the oldest man to fly in space and the first to make three flights. Ed White's parents were on hand for the launch.

Nixon Vows End To Tax Surcharge (Page 20B)

Weekend Edition

The News-Sentinel

'Sniffles' Schirra Shun Space TV Appearanc

GUARD OUSTS ARIAS THIRD TIME

Panama Rebels Tighten Grip

Astrona Plagued By Cold

Reds Mauled in Close-Quarter Fights Near Green Beret Camp

U.S. Concerned, Sec. Rusk Says

LBJ WINS, LOSES

Wally Schirra developed a cold overnight after the first day in orbit. And the first television show from space had to be delayed until later in the flight. The main objectives of Apollo 7 were to test out all of the many engineering systems aboard the Moonship and it had already been announced that if all went well Apollo 8 would head for the Moon in December. Everyone was keeping their fingers crossed. It is not outside the realm of speculation that the many pressures related to this first flight of the Apollo may have helped cause Wally's cold, coupled with the fact that the crew had to keep their bulky spacesuits on for the first 8 hours of the flight.

Amid sniffles and colds were many moments of levity during the first live TV transmissions from an American spaceship.

Apollo 7 spent more man hours in space than all the Soviet space flights combined up to that time. The mission featured the first live TV from a manned spacecraft. In a numerical odyssey Wally Schirra flew Mercury 5, Gemini 6 and Apollo 7. Apollo 7 carried a Lunar Module Pilot, but no Lunar Module.

Agnew To Stump Here Tuesday (Page 1B)

The News-Sentinel

CITY EDITION

U.S. Given First 'Live' Look at Apollo Crewmen

New Red Rockets Hit City

Army 'Racist'

TV Shows Spacemen In Action

REATS CITED

hird Teacher Strike ripsGotham Schools

On the 45th orbit of their 163 orbit, 11 day flight the cold-plagued crew made its first telecast. The show opened with Wally Schirra holding up a sign that read "Hello from the lovely Apollo room high atop everything." Another sign held up by Wally, who had once flashed a "Beat Army" sign in the window of his Gemini spacecraft, read "Keep those cards and letters coming in, folks." Back on Earth, Flight Director Eugene Kranz said: "We're ticking off every test like, gee whiz, like we've done it a thousand times and this is the first time. It's going great!"

Chamber Hears Tax Hike Forecast (Page 1C)

The News-Sentinel

CITY EDITION

Apollo 7 Capsule Lands Upside Down Nine Miles From Carrier; Trio Safe

Verbal Jabs Swapped by Candidates

Nixon Says Peace, Security Demand Humphrey Defeat

No Force In State On Time

Christmastide Flight Around Moon Likely; Ears Endure Plunge

28 MINUTES OF AGONY

Champagne Pops

During the flight Apollo 7 fired the Service Module engine eight times. This was a critical test, for without this engine no Moon Mission would be possible. On the first firing the engine jolted the Astronauts more than they expected and Schirra yelled "Yabadabadoo" like Fred Flintstone, the popular cartoon character of the day. The crew left their helmets off for the reentry so they could clear their cold-stuffed ears and avoid burst eardrums. In a space cold, mucous accumulates, filling the nasal passages and does not drain from the head because of the lack of gravity. There was an anxious moment at the end of the mission when splashdown occurred and the Apollo flipped over like a beached turtle. For 20 anxious minutes the world thought the worst until radio contact was finally established. Had the crew passed out because of the intense ear pain, or had the Apollo split in two and sunk after hitting a huge wave? Such was the speculation.

Earth from the Moon. Night is falling across the blue planet. These Apollo 8 photographs forever changed Man's concept of his small home.

APOLLO 8

A moment of relaxation for the Apollo 8 crew before setting off for the Moon. Kennedy Space Center crew quarters.

James A. Lovell, Jr.
Cleveland, Ohio

Frank Borman
Gary, Indiana

William A. Anders
Hong Kong

Series: Apollo
Date: December 21-27, 1968
Crew & Age: CDR. Frank Borman (40) Gemini VII
CMP. Jim Lovell (40) Gemini VII, Gemini XII
LMP. Bill Anders (35)

Distinction: First humans to journey to the Earth's Moon • "The Christmas Eve Bible Reading" • First pictures of Earth from Deep Space • New World Speed Record—24,200 MPH

Crew Insignia: Apollo 8 Patch, U.S. Flag, NASA Insignia

Spacecraft Name: Apollo Eight

"The design of the Apollo 8 patch was quite unique. Borman and I were in California working on our Apollo spacecraft when we got the word that our mission had been changed. We were going to take McDivitt's spacecraft and make a circumlunar flight around the moon.

On the way back to Houston the next evening, Frank was flying the airplane, and since I had nothing to do I sort of sketched out what I thought would be an appropriate patch. The shape of the patch symbolizes the Apollo spacecraft. The figure 8 signifies Apollo 8 and also the flight path we took to the moon and back. After I returned to Houston I gave my sketches to the NASA artist who made the final drawing.

By the time we started to fly Apollo 8 we wanted to name our spacecraft, but NASA said "no". The tentative name we had chosen was the "Columbiad" which was the name of the cannon in Jules Verne's book "From the Earth to the Moon." This flight and Verne's fictional story almost 100 years earlier had many similar aspects to it."

—Jim Lovell
Command Module Pilot

Both spacecraft carried three Astronauts.
Borman, Lovell and Anders in Apollo 8.
Barbicane, Nicholl and Ardan in Verne's Columbiad.
Both spacecraft were launched in December.
Both were launched in Florida.
Jules Verne's launch pad was only 100 miles west of Apollo 8's Cape Kennedy.

Entering the transfer van for the eight mile ride from the crew quarters out to the launch pad. Note the crew emblem on the door.

It's Not Too Late to Help Globe Santa

Boston Sunday Globe

'This flight will be duly recorded as man's first step away from his abode to another heavenly body'

'I can see the entire earth now... I can see Florida, Central America, all the way down to Argentina'

Man Leaves His Planet For Historic Moon Trip

APOLLO 8: THREE MEN CIRCLE THE MOON

150,000 Miles By Noon Today

The decision to "go for the Moon" on Apollo 8 had surfaced late in 1967 when George Low and Chris Kraft had pushed for the idea as a good way "to learn more about communications, navigation and thermal control in a deep space environment." In order to keep the idea secret for as long as possible they referred to the blossoming idea as "Sam's Budget Exercise." At the meeting where the final decision was made, Deke Slayton, the Astronaut's boss, said: "It is the only chance to get to the moon before the end of 1969." Right up until early December there was speculation that the Russians would once more try to steal America's thunder with a lunar fly-by mission of their own. The U.S.S.R. launch window was December 8. They did not use it. At 7:51 a.m. Saturday morning, December 21, 1968 Apollo 8 departed Earth. At 10:17 a.m. Mike Collins, in Mission Control, uttered the historic words never before heard, "All right, you are go for TLI (Translunar injection—the rocket firing toward the Moon)." You are "go" for escape velocity from cradle earth. Mike was originally scheduled to be on this historic flight, but a bone spur on his spine had to be operated on and his back-up Jim Lovell, climbed aboard instead.

The Evening Star NIGHT FINAL

Astronauts Circle Moon

Surface Looks Like Grayish 'Dirty Beach'

The Moon—about 70 miles away

The Earth—200,000 miles away

Red Attacks Mar Viet Truce

Pueblo Crew on Way Home; Bucher Doubts Seizure Plan

On Christmas Eve Apollo 8 went into lunar orbit, firing their engine on the far side of the Moon, out of voice contact with Earth. Had they perished? The world held its breath as an estimated 1 billion people in 64 countries listened in. Finally, Paul Haney in Mission Control happily announced: "We've got it! We've got it! Apollo 8 now in lunar orbit." Jim Lovell's first impression was that "It looks like plaster of paris, or sort of grayish beach sand." Bill Anders said it looked like dirty beach sand to him "with a lot of footprints." At about 7:30 a.m. they broadcast live television pictures of the lunar surface back to Earth.

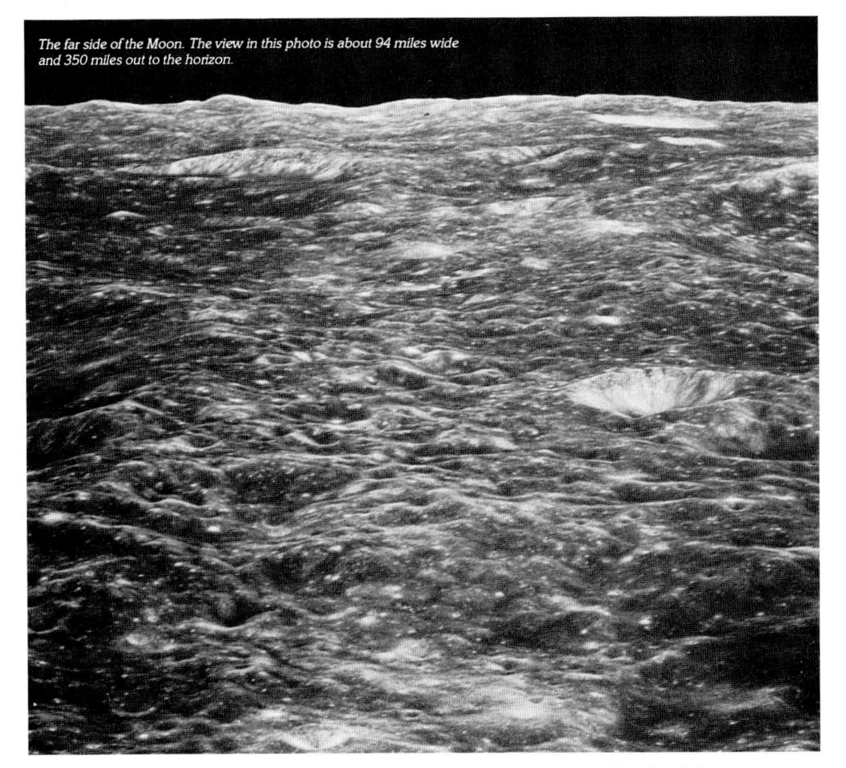

The far side of the Moon. The view in this photo is about 94 miles wide and 350 miles out to the horizon.

While orbiting the Moon the American Astronaut's named craters for the 8 American Astronauts who had died: Theodore C. Freeman, Elliot M. See, Jr., Charles A. Bassett II, Virgil I. Grissom, Roger B. Chaffee, Edward H. White II, Edward G. Givens Jr. and Clifton Curtis Williams.

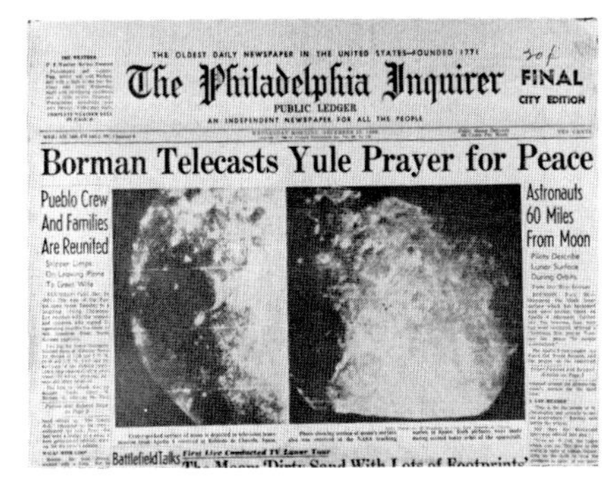

THE OLDEST DAILY NEWSPAPER IN THE UNITED STATES—FOUNDED 1771

The Philadelphia Inquirer

FINAL
CITY EDITION

PUBLIC LEDGER

AN INDEPENDENT NEWSPAPER FOR ALL THE PEOPLE

Borman Telecasts Yule Prayer for Peace

Pueblo Crew And Families Are Reunited

Astronauts 60 Miles From Moon

Battlefield Talks

None of us will ever forget the reading from lunar orbit that the three crewmen did from the first 10 verses of the Book of Genesis in the Bible—"In the beginning, God created the heavens and the Earth..." This they did on the 9th revolution. On the 3rd orbit Frank Borman, a lay reader at St. Christopher's Episcopal Church near the space center, offered a Prayer for Peace "to people everywhere." Both were very religious, moving moments and there were not too many dry eyes in homes around America when he and his crewmates were done. We were all very proud. They ended the Genesis reading with "Good night, good luck, a Merry Christmas and God bless all of you...all of you on the Good Earth." The author later produced a film with Jim Lovell about outdoor recreation on this planet titled "The Good Earth." It has been seen by millions on television and in the Nation's schools.

CHICAGO DAILY NEWS

Final Markets
Red Streak
Stocks Down

They're back!

'Moon is American cheese:' Borman

Man, 50, survives
Gift heart surgery here successful

Perfect Pacific landing

On the 10th orbit of the Moon, on Christmas Morning, the crew fired their main engine to begin the return to Earth. As they rounded the corner from the backside of the Moon, Jim Lovell radioed back to T.K. Mattingly at Cap Com, "Please be informed there is a Santa Claus." The trip back was uneventful, although many worried about the Moonship hitting the narrow reentry corridor. Too shallow an angle and they would skip off into outer space, lost forever. Too steep an angle and the Apollo would burn up in the fire of reentry into the atmosphere. Neither happened, and as they waited for recovery after splashdown, Borman quipped to Don Jones the helicopter commander of Madison, Wisconsin, who asked what the moon was made of... "It's not made of green cheese, it's made of American cheese."

McDIVITT SCOTT SCHWEICKART
USA
APOLLO IX

APOLLO 9 "Gumdrop & Spider"

James A. McDivitt
Chicago, Illinois

David R. Scott
San Antonio, Texas

Russell L. Schweickart
Neptune, New Jersey

Series: Apollo

Date: March 3-13, 1969

Crew & Age: CDR. James A. McDivitt (39) Gemini IV
CMP. David R. Scott (36) Gemini VIII
LMP. Russell L. Schweickhart (33)

Distinction: First Test of Lunar Module in Space • First Test of Portable Life Support System in Space • Rendezvous and Docking after 6 hour and 113 mile separation in Space.

Crew Insignia: Apollo 9 Patch, U.S. Flag, NASA Insignia

Command Module: Gumdrop

Lunar Module: Spider

"For Apollo IX, the mission patch was designed by Dave Scott, Rusty Schweickhart and myself. It was the first flight of the lunar module and the major objectives of our mission were to demonstrate the lunar module by itself and the lunar module and command module together. Consequently the design as it is shown. Also, the mission was called the "D" mission. There were "C", "D", "E", "F", and "G" missions...each with a certain number of objectives. Our mission was the "D" mission and the "D" in McDivitt had a red interior which signified the "D" mission.

The names for the vehicles were "Spider" and "Gumdrop." The Command Module looked like a gumdrop and the lunar module looked like a spider. This was the first mission in which the use of names for the spacecraft was again authorized. There was no way you could fly a mission with two spacecraft and call both of them by the same call sign. So we went to the names again and picked "Spider" and "Gumdrop"...not very glamourous, but they certainly fit the picture."

—Jim McDivitt
Commander

The Lunar Module, or "15 Ton Taxi" as it was sometimes called, cost $41 million dollars per copy. Its fuel weighed 3 times the LM itself. It was the first vehicle built to operate only in the vacuum of outer space. It had no reentry heatshield and could not be used for landings back to earth.

The first flight of the lunar module with Jim McDivitt and "Rusty" Schweickart at the controls. A daring episode in spaceflight history.

Dave Scott in the Apollo hatch. Part of the lunar module "Spider" is in the foreground.

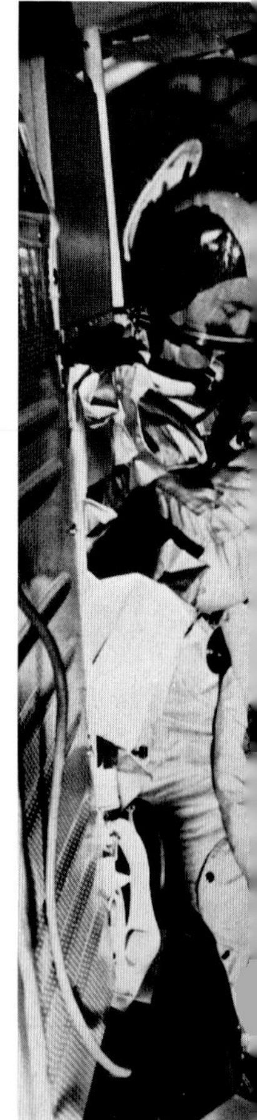

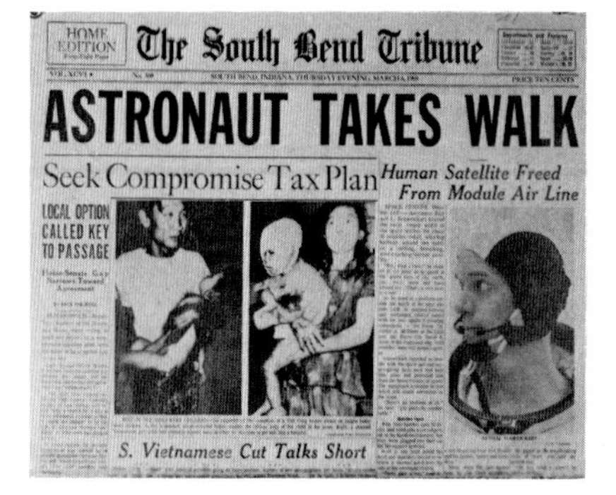

HOME EDITION

The South Bend Tribune

ASTRONAUT TAKES WALK

Seek Compromise Tax Plan

Human Satellite Freed From Module Air Line

LOCAL OPTION CALLED KEY TO PASSAGE

S. Vietnamese Cut Talks Short

CHICAGO DAILY NEWS

Markets Red Flash

2 solo in lunar 'spider'

$2,000 for 'fix'

Bribe try charged in convention case

Gold price

Hanoi's fist

Saltonstall

There had been no lunar module on either Apollo 7 or Apollo 8. This would be its first flight in space, a particularly critical test of this most important piece of hardware. If the LM didn't operate properly, there would be no landing on the Moon, at least not before our national goal of 1970. Rusty Schweickhart also had the task of testing out the spacesuit and portable life support system that would be worn on the Moon. It had to be tested in the vacuum of outer space. His original walk in space was shortened after he had vomited twice during the mission. It would be disastrous to have this happen while he was fully suited up and outside the spaceship. Rusty stood on the front porch of the LM "Spider" with his feet tucked into "Golden Slippers" to stay in place and Dave Scott stood in the open hatch of the Command Module. With the communications set-up, the Command Module was now "Gumdrop", the Lunar Module "Spider" and Rusty was "Red Rover" in honor of his red hair. His stomach held up fine.

When Jim McDivitt first fired the lunar module descent propulsion engine it "chugged noisily" at 20 percent throttle. He stopped throttling and waited. Was Spider a lemon? Within seconds the chugging stopped and he throttled up to 40% with no problems. The next time he tried the engine it worked perfectly. He had radioed back, "This really is an ungainly beast." One newspaper account said the spindly-legged craft looked like a "flying bedstead." Rusty and Jim in Spider could not return to Earth if anything happened in their maneuvers 100 miles from the Command Module. Dave Scott in Gumdrop was prepared to come out and rescue them if needed.

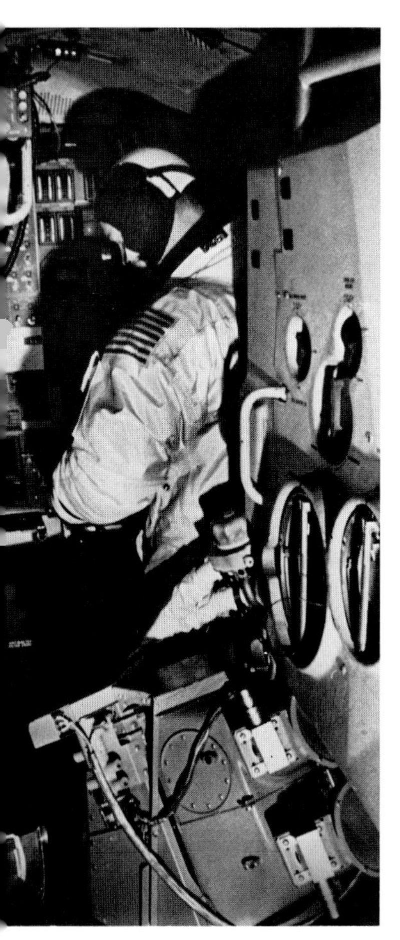

Dave Scott flies in formation in Gumdrop.

Jim McDivitt (left) and "Rusty" Schweickart inside the lunar module. It was a compact flying machine, indeed.

Chicago Tribune

DOCK MOONCRAFT IN SPACE

61st Auto Show Opens

TENSE LINKUP TAKES CREW 20 MINUTES

Success!

REPAY STATE 7 MILLION IN AID, CHA TOLD

Spectators Get Look at High Styles

Ogilvie Assails Fund 'Grab'

Last Hurdle for Lunar Landing

TOGETHERNESS

Death of Uris Bride Suicide, Jury Rules

THE INDIANAPOLIS STAR

U.S. 4 MONTHS FROM MOON

Senate Passes Nuclear Curb Treaty

Influenced By ABM Opposition

Spacemen Splash Then Salute

Apollo-9 Lands Safely

Uni-Gov Council Organizational Effort Due Soon

Israeli, Egyptian Gunners Duel

About a year before Apollo 9 there had been serious concerns raised about flying the Lunar Module once its lower stage had been dropped off. Since it had to serve as its own launch pad on the Moon, the lunar module was made in two sections. The top section held the small crew cabin, the lower contained the descent engine and fuel that would land them on the Moon. But we had to know if the two sections would be successfully separated and flown in a vacuum. The other concern was with the redocking of the Command Module and Lunar Module. Would "jack-knifing" occur and damage one of the spacecraft? So when Jim McDivitt fired the pyrotechnics to drop off the descent section he didn't really know what to expect and he reacted with the comment that it was "sort of a kick in the fanny... but it went all right." After being separated for more than 6 hours the two vehicles were safely docked and that part of the grand adventure was over. An exhausted McDivitt said that he was "going to take three days off. That wasn't a docking, that was an eye test."

At the end of 10 days, 1 hour and 1 minute Apollo 9 splashed down safely in the Atlantic. Lunar Module Development Leaders Carroll Bolender of the Manned Spacecraft Center and Llewellyn Evans of Grumman Aircraft were given the NASA Exceptional Service Medal and the NASA Public Service Award. They and all of the Americans who had designed and built the incredible flying machine "Spider" shared in the success of Apollo 9. But there was the one remaining problem, the chugging of the Spider's engines caused by helium ingestion would have to be solved before the LM was judged 100% ready for an actual landfall on the Moon.

Earthrise from Apollo 10.

APOLLO 10 "Snoopy and Charlie Brown"

Eugene A. Cernan
Chicago, Illinois

John W. Young
San Francisco, California

Thomas P. Stafford
Weatherford, Oklahoma

Series Apollo
Date: May 18-26, 1969
Crew & Age: CDR. Tom Stafford (38) Gemini VI, Gemini IX
CMP. John Young (38) Gemini III, Gemini X
LMP. Gene Cernan (35) Gemini IX
Distinction: "The Dress Rehearsal" • First use of Color TV in Space • Tested Lunar Module in Lunar Orbit

Crew Insignia: Apollo 10 Patch, U.S. Flag, NASA Insignia
Command Module: Charlie Brown
Lunar Module: Snoopy

"The Apollo X patch was based more on the mechanics and goals of the mission than the philosophy of the space program or of the astronauts flying this mission. The big Roman Numeral X is supposed to give the three dimensional effect of sitting on the surface of the moon. The command module "Charlie Brown" is in orbit as the lunar module "Snoopy" returns from its low pass over the surface of the moon. The patch was basically designed by the crew, primarily Young and Cernan, with a great deal of help from the artists at North American/Rockwell. North American and Grumman eventually were the producers and primary suppliers of the patch. Although Apollo X did not land on the moon, it was meant to indicate through the prominence of the Roman Numeral X that Apollo X had left its permanent impression.

The lunar module lent itself to the name "Snoopy" (after Charles Schulz' popular cartoon strip of the day) because that is really what it was doing on Apollo X, snooping around the lunar surface.

The name "Charlie Brown" for the command module seemed appropriate as the ever present companion and guardian of "Snoopy".

—Gene Cernan
Lunar Module Pilot

Tom Stafford pats "Snoopy" on the head on the way to the transfer van. Snoopy is being held by Jayme Flowers, Gordon Cooper's secretary at the time.

The Apollo 10 crew with their Mission Patch and mascot "Snoopy" at a Press Conference on April 26, 1969 in Houston.

The Miami Herald

LM Carries Out Landing Dry-Run

Friends Out As Appointees Of President

2 Men Scan Moon At 'Airliner Height'

'Magnificent!' Cernan Shouts

Where Snag Developed

Today's Schedule

Chicago Tribune

HOME

PASS MOON AT 50,000 FEET

'WE IS DOWN AMONG 'EM,' SAYS CERNAN

Minor Problems Overcome

Talks Go On as Teachers Strike

Ultimatum Ends Sit-In at Circle U.

MUCH LEFT TO SETTLE, SAYS UNION

Some of Staff Show Up

Apollo 10 could have landed on the Moon, and would have, according to the official NASA history "Chariots for Apollo", if the flight had been scheduled for the last few weeks of John Kennedy's Decade. As it was, however, we still had time to test the unproven lunar module guidance and navigation system in the uneven gravity fields of the Moon, and near enough to the Command Module so that it could rescue the men in the lander if that became necessary. So Apollo 10 got hung with the term, "dress rehearsal", if any trip out to the Moon and back could be looked at in such a trite way. But, what the heck, we'd already been out there once. On May 18, 1969 America's first all-veteran Astronaut crew launched from Pad 39B, the only launch from that facility that would occur in the Moon Program. On hand were King Baudoin and Queen Fabiola of the Belgians who had been flown to the launch aboard the President's Air Force One. The launch into Earth orbit was very rough because of the Pogo effect of the Saturn, and the crew wondered seriously if the machine would hold together so they could continue on their mission. When they finally arrived at the Moon Stafford radioed back, "Houston, tell Earth we have arrived."

On the morning of the 22nd, Mission Control played "The Best is Yet To Come," and sounded reveille only to discover that the crew was already awake and raring to go. After a scare with a misaligned docking mechanism, the lunar module "Snoopy" left the command module "Charlie Brown" and swooped down toward the lunar surface. Tom Stafford and Gene Cernan were practicing DOI (Descent Orbit Insertion) for the first time. "We is go and we is down among 'em" was the way Gene described the view from 50,000 feet, "We just saw Earth rise, and its got to be magnificent." Stafford chimed in with, "There's enough boulders around here to fill Galveston Bay."

The Command Module "Charlie Brown" in lunar orbit.

"Charlie Brown" during reentry into the Earth's atmosphere across the dark skies of the South Pacific.

John Young and Gene Cernan inspect their charred Command Module aboard the recovery ship U.S.S. Princeton.

THE PLAIN DEALER

OHIO'S LARGEST NEWSPAPER

ALL-OHIO EDITION

2 ASTRONAUTS IN 'SNOOPY' HEAD FOR 'CHARLIE BROWN'

Craft's Spin Is Stopped

APOLLO 10

Gilberts Still for Busing Plan | Broncos Announce '69 Schedule | Mary Bond Lives in a Crate

REPORT ON CU TURMOIL

Regents Urged to Ban SDS

THE DENVER POST

HOME EDITION

The Voice of the Rocky Mountain Empire

GOOD NEWS TODAY

Lunar Landing Feasible

Rulison Test To Check On Disarm Idea

Crew Flies Module to Moon's Edge

Suddenly, as Stafford and Cernan in "Snoopy" whipped low over the Moon the Lunar Module threw a fit and began lurching wildly about. Gene Cernan's gut cry of "@★!!@×†§★!!" was followed by Stafford's that they were close to Gimbal Lock, that the engine had swiveled over to a stop and stuck. Luckily that had not happened and Stafford took over manual control as "Snoopy" continued his wild "gyrations across the lunar sky." Finally, after a scary three minutes they brought the lander under control following jettison of the lower half of "Snoopy", the descent stage. The trouble was traced to a switch in the wrong position, nearly a fatal mistake. "Snoopy" had been gone for more than eight hours when he finally rejoined "Charlie Brown."

After 31 orbits of the Moon, Apollo 10 headed home, streaking through the early morning darkness on May 26, like a shooting star, the splash down near American Samoa. The Moon was within our grasp. If the "creek didn't rise" and all went well, we would try to fulfill John Kennedy's Goal of "Landing a Man on the Moon and returning him safely to Earth" on our next flight, Apollo 11.

Neil Armstrong's footprint in the lunar soil.

APOLLO 11 "Columbia and Eagle"

Neil A. Armstrong
Wapakoneta, Ohio

Michael Collins
Rome, Italy

Dr. Edwin E. Aldrin, Jr.
Montclair, New Jersey

Series: Apollo
Date: July 16-24, 1969
Crew & Age: CDR. Neil Armstrong (38) Gemini VIII
CMP. Mike Collins (38) Gemini X
LMP. Buzz Aldrin (39) Gemini XII
Distinction: First Manned Landing on the Earth's Moon • Established First Moon Base • First Humans to Walk on the Moon • Fulfilled President Kennedy's National Goal

Crew Insignia: Apollo 11 Patch, U.S. Flag, NASA Insignia
Command Module: "Columbia"
Lunar Module: "Eagle"
Moon Base: "Tranquility Base"

"We needed something simple, yet something that unmistakedly said peaceful lunar landing by the United States. Jim Lovell, Neil's back-up, introduced an American Eagle into the conversation. Of course! What better symbol. At home I skimmed through my library and finally found what I wanted in a National Geographic book on birds: a bald eagle, landing gear extended, wings partially folded, coming in for a landing. I traced it on a piece of tissue paper and sketched in a oblique view of a pockmarked lunar surface. I added a small earth in the background and drew the sunshine coming from the wrong direction, so that to this day our official insignia shows the earth over the lunar horizon like when it should really look like . I also penciled in "Apollo" around the top of my circular design and ELEVEN around the bottom. Neil didn't like the ELEVEN because it wouldn't be understandable to foreigners, so after trying XI and 11, we settled on the latter and put Apollo 11 around the top. Jim Lovell and I both agreed that the Eagle alone really didn't convey the entire message we wanted. The Americans were about to land, but so what? Tom Wilson, our simulator instructor, overheard us and piped up, why not an olive branch as a symbol of our peaceful expedition? Beautiful, where do eagles carry olive branches? In their beaks, naturally. So, I sketched one in and after a few discussions with Neil and Buzz over color schemes, we were ready to go to press. Stan Jacobsen in Houston assigned an artist on his staff, James R. Cooper, to finalize the design for us.

NASA disapproved the original design as too hostile and warlike because of the Eagle's powerful talons extended stiffly below him. What to do? A gear-up approach was unthinkable. Perhaps the talons could be relaxed and softened up. Then someone had a brainstorm; just transfer the olive branch from beak to claw and all menace disappeared.

The choice of an Eagle as a motif for the landing led swiftly to naming the landing craft Eagle. One day I was chatting long distance with Julian Scheer, NASA Public Affairs, and he suggested the name Columbia. It sounded a bit pompous to me. But Columbia did have a lot of things going for it—the close similarity of Jules Verne's mythical moonship cannon, the Columbiad, and the close relationship between the word "Columbia" and our national origins. Columbia had almost become the name of our country. Finally, the lyrics "Columbia, the Gem of the Ocean" kept popping into my mind and they argued well for the recovery of a spacecraft which hopefully would float on the ocean.

Since Neil and Buzz had no objections and since I couldn't come up with anything better, Columbia it was, and I think not a bad choice."

—Mike Collins
Command Module Pilot

Excerpts and edited from "Carrying the Fire" by Michael Collins, An Astronaut's Journeys, foreworded by Charles A. Lindbergh. Ballantine Books. Copyright 1974 by Michael Collins.

The Apollo 11 crew left a medallion of the Apollo 1 patch on the moon to commemorate the fallen crew of Apollo 1—Gus Grissom, Ed White and Roger Chaffee.

Apollo 11 Landing area—The Sea of Tranquility
Touchdown—4:10 p.m. EDT Sunday, July 20, 1969
Coordinates—0° 41" 15 sec. North Latitude 23° 25" 45 sec. East Longitude
Man Steps on Moon—10:56 p.m. EDT, Sunday, July 20, 1969
Lift-Off from Moon—1:54 p.m. EDT, Monday, July 21, 1969

Neil Armstrong in the Lunar Landing Training Vehicle (LLTV). A year before his Apollo 11 Mission, his life was saved by the ejection seat in his LLTV when it suddenly went out-of-control. The seat was made by the Weber Aircraft Division of Kidde, Inc., the author's employer.

Eshkol dies: Israel power fight
Story below in Column 1

CHICAGO DAILY NEWS

Mild

Markets Red Flash

Daily News exclusive

1st man on moon picked

Heart attack
Eshkol, 73, dies; Israel power fight

South Side CHA guard shot dead

Air Force's Col. Aldrin given role

This front page erroneously identified Buzz Aldrin as being selected to be the First Man To Walk On The Moon!

Orlando Sentinel
'Tis a Privilege to Live in Central Florida

Eyes Of The World Watching Cape

Apollo 11 Ready To Sail Into History

Optimism Prevails On Success

Area Roads Jammed By Car Influx

Astronauts 'Top Shape' Physically

Luna To Hit, Run Home?

Apollo Index

My family and I joined over a million others from Titusville to Cocoa Beach to watch the historic launch. It was a warm summer day.

The Apollo 11 plaque still on the Moon.

"There was a slight change in the plaque we were to leave on the moon—a change made by President Nixon. The original read, "We come in peace for all mankind." Mr. Nixon changed "come" to "came," a decision we all supported.

Neil and I had another decision to make: what to call—for communications purposes—the exact place on the moon where we would land. It would be somewhat similar to a radio call sign, but we wanted to give it added significance. Moon One? Base Camp? Moon Base? When we made our choice, we told only Charlie Duke, who would be our Capsule Communicator back in Houston, and who we felt should know the exact name in case transmission was garbled. I cannot remember which of us originated the selection, but once we had thought it over it was an obvious choice. We were landing in an area known as the Sea of Tranquility; we would call our landing site Tranquility Base."

—Edwin E. "Buzz" Aldrin, Jr.
Lunar Module Pilot

From "Return to Earth" by Colonel Edwin E. "Buzz" Aldrin, Jr. with Wayne Warga. Random House.

THE ATLANTA CONSTITUTION

ACTION LINE

Soviet Luna Racing To Reach Moon as Apollo Takes Off

Called No Hazard To U.S. Astronauts

THE JOURNAL-GAZETTE

Weather For Launch To Be Good

Russ Send Unmanned Craft To Moon; May Be Lunar Rock Scoop

Might Be Hazard For Apollo 11

600 Marines Sail Away From Viet

Health Stand Hit

Pickets, Card Burner Upset AMA Meeting

Trio Flee Jail, Hold Hostages

Festival Caught On

Sir Bernard Lovell in Great Britain at Joddrell Bank Observatory said LUNA 15 appeared to be racing to the Moon to beat America to the Moon rocks, even as Apollo 11 sat ready on the pad.

That the Russians were attempting to steal the limelight away by upstaging the U.S. was joined by the news leaked that some in the U.S. Space Program feared that the Moon Rocks and dust would burst into flames once brought back on board the Lunar Module by the Astronauts.

"Buzz" Aldrin steps onto the Moon. Photo taken by Neil Armstrong with his Hasselblad 500 EL camera. Neil carried the only camera, so there are no good pictures of him other than those from the TV screen, an ironic twist to the Apollo 11 Mission. Neither crewman remembered to take Neil's picture.

Neil Armstrong (left) and "Buzz" Aldrin reading from the plaque on the landing strut of the LM Eagle. It is still on the Moon's surface. "Here men from the Planet Earth first set foot upon the Moon. July 1969, A.D. We came in Peace for all Mankind."

"Buzz" Aldrin and Old Glory on the Moon at Tranquility Base.

Neil Armstrong inside Eagle at Tranquility Base. The crew's historic moonwalk was over, as can be seen by the relief reflected in Neil's face.

"Buzz" Aldrin inside Eagle at Tranquility Base. "Buzz" celebrated Communion while on the Moon's surface in a private moment of remembrance.

This olive branch, traditional symbol of peace, was left on the Moon by the crew of Apollo 11.

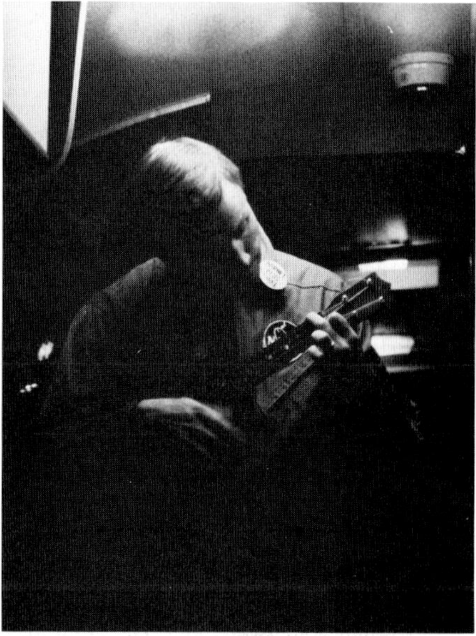

A musician from his high school days, Neil relaxes in the Mobile Quarantine trailer by playing his ukulele while still aboard the recovery ship, the U.S.S. Hornet.

The Apollo 11 crew inside the Mobile Quarantine Facility after their return to Earth. This was to guard against bringing any foreign organisms back to Earth from the Moon.

COUNT DOWN
First men land on moon

Good Evening

Wapakoneta Daily News

NEIL STEPS ON THE MOON

"All people one," Nixon tells pair

Navy beefs up armada in Pacific

Armstrong, Aldrin set for hazardous return

Mom, dad worries lighter after walk

Neil's Hometown Newspaper

Chicago Tribune

FINAL MARKETS ★ HOME

3 MOON HEROES WIN CITY

Million Line 7-Mile Parade Route

RUSS, CHINA FORCES FIGHT OVER BORDER

Heavy Casualties Told by Peking

Chicago Has Greatest Day

At the State Dinner in Los Angeles, President Nixon presented posthumous medals to the three dead Apollo 1 Astronauts, Gus Grissom, Ed White, and Roger Chaffee saying "We have the privilege sometimes of standing on the shoulders of the giants who have gone before us." Steve Bales, a flight control officer who gave the "Okay to land" signal when their computer almost forced Eagle to abort its landing, was also honored.

Chicago Tribune

MOON FINAL

MONDAY, JULY 21, 1969

FLY U.S. FLAG ON THE MOON

First Picture on the Back Page

"Tranquility Base here. The Eagle Has Landed."

The Miami Herald EXTRA

MAN WALKS ON THE MOON

'A Small Step for Man, a Leap for Mankind'

Spacemen Explore For Two Hours, Return to Eagle

"That's One Small Step for Man, One Giant Leap for Mankind."

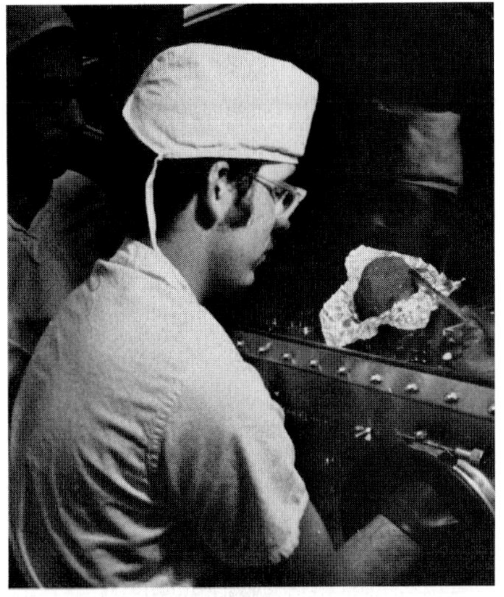

Working with the Moon rocks in a germ-free environment at the LRL in Houston.

Breccia, an Apollo 11 Moon rock.

"The Board of Directors." These were the men responsible for "making it happen." Left to right: Dr. George E. Mueller, NASA Headquarters; General Samuel C. Phillips, Director, Apollo Program; Dr. Kurt H. Debus, Director, Kennedy Space Center; Dr. Robert R. Gilruth, Director, Manned Spacecraft Center; and Dr. Wernher von Braun, Director, Marshall Space Flight Center.

APOLLO XII
Conrad · Gordon · Bean

APOLLO 12 "Intrepid and Yankee Clipper"

Charles Conrad, Jr.
Philadelphia, Pa.

Richard F. Gordon, Jr.
Seattle, Washington

Alan L. Bean
Wheeler, Texas

Series Apollo
Date: November 14-24, 1969
Crew & Age: CDR. Pete Conrad (39) Gemini V, Gemini XI
CMP. Dick Gordon (40) Gemini XI
LMP. Al Bean (37)
Distinction: Pinpoint Landing 535 feet from Surveyor III crater • 1st Moonwalk 3 hours 56 minutes, 2nd 3 hrs. 49 min. • Set-up sophisticated Geophysical Station • Initiated a Lunar Nuclear Power Station

Crew Insignia: Apollo 12 Patch, U.S. Flag, NASA Insignia
Command Module: Yankee Clipper
Lunar Module: Intrepid

"The patch was designed by Pete, Dick and I with the help of about 10 other people around the contractor area at the Cape. The real breakthrough came in our effort to try to duplicate our landing site on the moon when a couple of engineers came to me and said they thought they could duplicate it exactly. They came back a few days later and it did look just perfect—all the craters the proper size, shape and the lighting just perfect. It turned out the technique they had used was to get a relief globe of the moon that was in the library and light it properly and then take photographs with a Polaroid at different distances until they got one that had just the right curvature that we wanted on the patch. We selected the blue and gold because they are Navy colors and all of us were in the Navy. The ship was patterned in a way after a Navy ship. I went to the library and picked up some clipper ship pictures, gave them to an artist who was working at Patrick at the time. He was the one who drew it. After looking it over, we realized there were too many sets of sails outboard of the hull, so we asked him to redesign it. The ship had been one similar to the one Jason and the Argonauts had used to search for the Golden Fleece and we didn't feel that was American enough, where we felt that the clipper ship was definitely an American symbol."

—Al Bean
Lunar Module Pilot
4th Man on the Moon

Apollo 12 landed just 535 feet from Surveyor III as can be seen in this remarkable photograph. An amazing pinpoint landing thanks to Dr. William Lear and Ewen A. Whitaker. Dr. Lear did the computer work and Whitaker located precisely where Surveyor was on the Moon's surface.

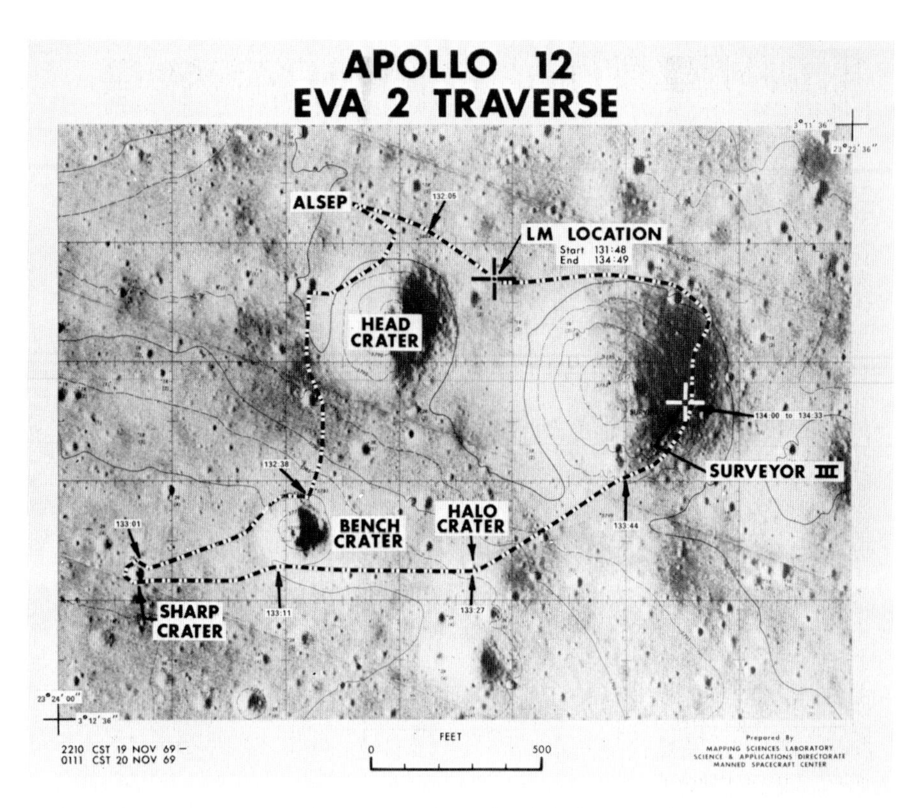

The following excerpt from "First on the Moon" by Little, Brown & Company:

"There were four stars on the patch of Apollo 12, one for each of the crewmen and one for C.C. Williams who would have been the lunar module pilot on this flight had he lived. Al Bean had replaced C.C. as Lunar Module Pilot, and it had been his idea to put the extra star there."

C.C. Williams died October 5, 1967 in the crash of his T-38 jet near Tallahassee, Florida.

Apollo 12 Landing area—The Ocean of Storms, (955 miles west of Apollo 11) Touchdown—1:54:35 a.m. EST, November 19, 1969 Coordinates—latitude 3° 11′51″ S. longitude 23° 23′8″ W Lift-off from Moon—9.25 a.m., November 20, 1969

Ruling Invalidates City Remapping (Page 1C)

The News-Sentinel

Lightning Fails To Daunt Apollo 12's Lift Off For Moon; Ship Orbits Earth

Saigon Pilots Err, Attack

APOLLO 12

VIOLENCE MISSING

Silent Marchers

On November 14, 1969 President Nixon became the first chief executive to watch a manned space launch, however, seconds after launch the huge Apollo 12 stack disappeared into the clouds. Suddenly, lightning flashed around the Moonship, critically vulnerable at this stage of lift-off. Apollo 12 lost its platform and there was grave concern for its safety. Finally, after checking everything out carefully, the crew was given approval for the TLI (Translunar Injection) burn outward bound to the Moon.

Chicago Tribune

FINAL

WE DO IT AGAIN

Apollo 11 had missed its landing site by about 4 miles, acceptable on such a historic first mission, but not suitable for future missions to specific areas of the Moon. Dr. William Lear of TRW in Los Angeles worked out a new computer procedure for Conrad, Bean and Gordon to use. With it they successfully landed within 535 feet of the Surveyor craft that had landed 2½ years before. They touched down at 1:54 a.m. in the Ocean of the Storms, an apt target considering their scary storm launch.

An Apollo 12 crewman with a container of lunar soil. Note the checklist on his sleeve. Beginning with Apollo 14, the Commander wore red stripes on his arms so that NASA could distinguish who was who in photographs.

THE INDIANAPOLIS STAR

ASTRONAUTS HAVE A 'BALL'

U.S. Battle Casualties In Viet Soar

Reds' Push Believed Near Peak

"Why, They're Just Plain Folks, Like Us!"

Pete And Al Like Tots Let Loose In Toy Store

Proud Wives

Moonmen Whistle At Work

The Evening Star — NIGHT FINAL

Apollo Crew Explores Moon, Spots Surveyor and 'Mounds'

Senate Approves Lottery Draft Plan, Sends It to Nixon

Exuberant Pair Takes Walk After Pinpoint Landing

Adjectives (Amazing) Really Fly

'Right Down the Middle'

As Pete Conrad reached the bottom of the lunar lander's ladder and stepped on to the Moon he said, "Whoopie, man, that may have been a small step for Neil, but that's a long one for me. I'm going to step off the pad. Right, Up. Oh, is that soft. Hey, that's neat. I don't sink in too far. I'll try a little—boy, that sun's bright. That's just like somebody shining a spotlight in your eyes. I can walk pretty well, Al, but I've got to take it easy and watch what I'm doing. Boy, you'll never believe it! Guess what I see sitting on the side of the crater. The old Surveyor!" "The old Surveyor, yes, sir," was Bean's reply. The only snafu occurred when the television camera beaming pictures back to Earth was pointed at the sun and ruined. For the rest of the Moon mission we had to depend upon voice descriptions of what the crew was doing and seeing.

The two Moon walkers visited the Surveyor, examined it and snipped parts of it off to bring back to Earth for analysis. The first EVA (Extravehicular Activity, "Moonwalk" in ordinary language) lasted 4 hours, the second 3 hours and 46 minutes. They deployed the first Apollo lunar surface experiments package, took core samples and documented samples of rocks and soil. There was nothing stuffy about these two, they laughed and chuckled and just plain enjoyed their time on the Moon and so did we back here on Earth who listened in. It was like listening to suppertime radio serials back in the 1940's, we had to imagine what it looked like and that made it all that more fun.

"Home, Sweet Home"

APOLLO 13 "Odyssey and Aquarius"

James A. Lovell, Jr. *Cleveland, Ohio* — *John L. Swigert, Jr.* *Denver, Colorado* — *Fred W. Haise, Jr.* *Biloxi, Mississippi*

Series:	Apollo
Date:	April 11-17, 1970
Crew & Age:	CDR. Jim Lovell (42) Gemini VII, Gemini XII, Apollo 8 John L. "Jack" Swigert, Jr. (38) Fred W. Haise, Jr. (36)
Distinction:	Deep Space Abort, safely returned to Earth • Lovell became first man to fly twice to the Moon • First Mission failure in 22 U.S. Manned flights • Lunar Module used as lifeboat.
Crew Insignia:	Apollo 13 Patch, U.S. Flag, NASA Insignia
Command Module:	Odyssey
Lunar Module:	Aquarius

"By the time we came to Apollo 13, I wanted to do something a bit different from the patriotic patches which symbolized the previous Apollo flights. We started out designing this patch with the idea of the mythical god, Apollo, driving his chariot across the sky and dragging the sun with it. We gave this idea to an artist in New York City named Lumen Winter; and he eventually came up with the three horse design which symbolized the Apollo, but also included the earth and the moon.

"We decided to eliminate the names on the front face of the patch and instead put in the Latin "Ex Luna, Scientia" or "From the Moon, Knowledge." It is interesting to note that Lumen Winter, prior to making the patch for us, made a large wall mural of horses crossing the sky with the earth below which is prominently displayed at the St. Regis Hotel in New York City. The horses are very similar to the ones on our patch. However, in the mural he has a fourth horse which has fallen behind and that, ironically, could have been Ken Mattingly who was replaced just before our flight for fear he was going to catch the measles.

"We called the Command Module "Odyssey" because of the long voyage. It was only after I returned that I discovered the meaning of Odyssey is "a long voyage with many changes of fortune." How true for "13".

"The Lunar Module was named Aquarius. Contrary to popular belief, it was not named after the song in the play, "Hair", but after the Egyptian God, Aquarius. She was symbolized as a water carrier who brought fertility and, therefore, life and knowledge to the Nile Valley; and we hoped our Lunar Module, Aquarius, would bring life back from the moon."

—Jim Lovell
Commander—Apollo 13

The "mailbox" that kept the Astronauts from dying of carbon dioxide during the long crisis. Built of duct tape and "spare parts" it drew carbon dioxide from the cabin through this lithium hydroxide cannister and expelled it back into the LM as pure oxygen.

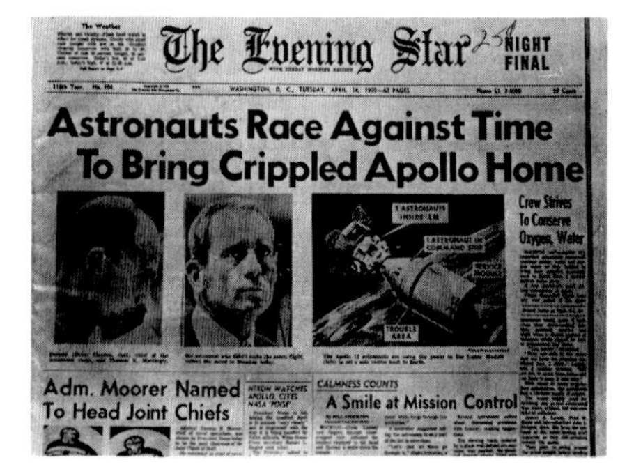

The Evening Star

NIGHT FINAL

Astronauts Race Against Time To Bring Crippled Apollo Home

Crew Strives To Conserve Oxygen, Water

Adm. Moorer Named To Head Joint Chiefs

CALMNESS COUNTS

A Smile at Mission Control

Chicago Tribune

IRON-NERVED CREW READY FOR SWING AROUND MOON

SLASH CTA's SUBSIDY BID TO WIN VOTES

Splashdown Due Friday

Apollo 13 shuddered when Oxygen Tank #2 blew up in the Service Module behind the crew's backs. Oxygen, electricity, light and water were lost and the 3 Astronauts had to use the tiny Lunar Module as a lifeboat. Apollo 13 was launched at 13:13 p.m. Houston Time and aborted shortly after 10:13 P.M. on April 13, 1970. Lost was approximately 13 Million Dollars worth of scientific equipment.

Jim Lovell has called the "Pericynthian plus 2" platform alignment, on Tuesday afternoon enroute to the Moon after the explosion, the biggest "heart-stopper" of the flight. Yet it went unreported in the Press at the time. Had this critical maneuver failed, they would have been marooned in space.

President Richard M. Nixon welcomes the crew back to Earth.

THE INDIANAPOLIS STAR

THURSDAY, APRIL 16, 1970

APOLLO COURSE CORRECTION MATTER OF LIFE OR DEATH

22 Yanks Killed In N. Viet Attack, Booby Trap Blast

Balky Spacecraft Could Miss Earth, Doom 3-Man Crew

The Press did, however, play up another course adjustment on Thursday, as seen in this headline. The 3 Astronauts were wet, cold and worried in their cramped LM quarters. The temperature dropped to 38°, water condensed on the cabin walls and frost formed on the inside of the windows. Designed for a 45-hour lifetime, they had to stretch the LM's reserves to 90 hours.

Cool

CHICAGO DAILY NEWS

The Complete Evening Newspaper

Markets Red Flash

Sizzling finish!

Drop 'dead' module, see blast hole

The crew jettisoned the blast-gutted Service Module just hours before reentry to Earth. They had towed it along for 300,000 miles because its bulk protected the Command Module's heat shield from the intense cold of space.

Alan Shepard (left) and Ed Mitchell during pre-flight training in Arizona. The two are at work at their MET (Modular Equipment Transporter) that they would later pull along on their moonwalks.

APOLLO 14 "Kitty Hawk and Antares"

Stuart A. Roosa
Durango, Colorado

Alan B. Shepard, Jr.
East Derry, New Hampshire

Dr. Edgar D. Mitchell
Hereford, Texas

Series: Apollo
Date: January 31-February 9, 1971
Crew & Age: CDR. Alan Shepard (47)
Mercury-Freedom 7
CMP. Stuart A. Roosa (37)
LMP. Edgar D. Mitchell (40)
Distinction: Third Lunar Landing, in the Fra Mauro region • Ed Mitchell experiments with ESP • Al Shepard hits two golf balls on the moon • Medical & manufacturing experiments in space • 95 lbs. of geologic specimens brought back from moon.
Crew Insignia: Apollo 14 Patch, U.S. Flag, NASA Insignia
Command Module: Kitty Hawk
Lunar Module: Antares

"Our crew wanted an aesthetically pleasing design with simplicity and style. We chose for the Earth and Moon to be portrayed and for the Astronaut symbol (the shooting star) to represent the crew. We went through a large number of tests with different shades of color for the planets and for the background before selecting the final design.

The Apollo 14 patch decision was a joint effort of Al, Stu Roosa and myself with the art work being a product of the NASA Art Department at Kennedy Space Center. Jean Bealieu was the artist who created the master rendering of the design."

—Ed Mitchell
Lunar Module Pilot
6th Man to Walk on the Moon

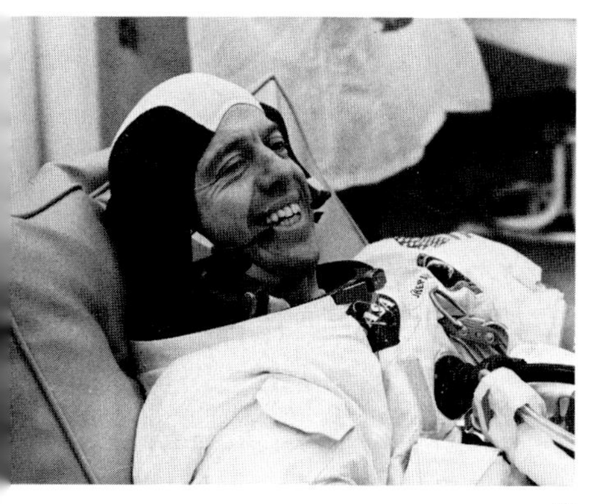

Al Shepard waited a long time for this pre-launch moment and underwent a secret ear operation to make it come true.

Antares at Fra Mauro. Latitude 3°40′24″ S., Longitude 17°27′55″ W.

"I was the one who named the command module "Kitty Hawk". I did it to honor the place where it all began with the *Wright Brothers* and felt this name was most appropriate. Ed named the lunar module "Antares." His logic was that Antares was the most visible "landmark" as they pitched over during the powered descent to the lunar surface. (Antares is a red star in the *Scorpius constellation.*) It is a very difficult chore to name a spacecraft. Ed and I spent hours trying to find two names that would be coordinated. Finally I told Ed, "You name the lunar module anything you want, but I'm going with Kitty Hawk."

Stuart Roosa
Command Module Pilot

THE PLAIN DEALER

OHIO'S LARGEST NEWSPAPER

APOLLO 14 LICKS LINKUP PERIL

Docking Works on 5th Try

Riot Follows Rally in L.A.; 1 Man Dies

Apollo 14 also had its share of glitches. To begin with, it left Earth 40 minutes late because of a light rainstorm in the area. As a matter of fact it was raining 3 miles away from the Pad at lift-off at the VIP viewing area where the author was standing and where Vice President Spiro T. Agnew, Neil Armstrong, Prince Juan Carlos and Princess Sophie de Borbon of Spain watched the launch. Four times Stu Roosa tried to dock with the lunar module, four times the capture latches failed to work properly. Finally, on try number five the two vehicles were mated, even as the crew was considering "going outside" to try to solve the problem. They were already 20,700 miles out from Earth. Had this not been resolved, no Moon landing would have been possible and with two Mission failures in a row, NASA may have been out of the Moon Exploration business.

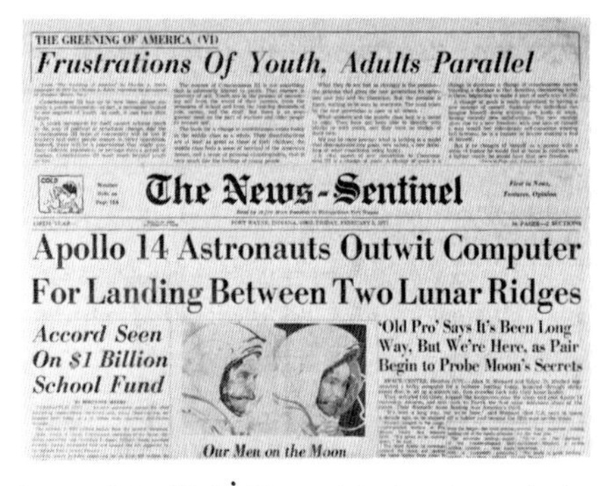

THE GREENING OF AMERICA (VI)

Frustrations Of Youth, Adults Parallel

The News-Sentinel

Apollo 14 Astronauts Outwit Computer For Landing Between Two Lunar Ridges

Accord Seen On $1 Billion School Fund

'Old Pro' Says It's Been Long Way, But We're Here, as Pair Begin to Probe Moon's Secrets

Our Men on the Moon

During revolution #12 of the Moon an electronic spook cropped up in the lander computer that would have caused the Antares to abort its landing try while in its Programs 63, 64 and 66 that were to be used for powered descent and landing on the Moon. The problem was probably caused by contamination in the abort switch. Massachusetts Institute of Technology, the prime contractor on the Apollo Guidance System (and Ed Mitchell's Alma Mater) worked out a solution to disable Program 70, the abort program. Al Shepard, age 47, brought the lunar module down to within 130 feet of the target. "It's been a long way, but we're here," said Shepard, America's first man in space 10 years before aboard the tiny Freedom 7.

The meandering tracks of men and their rickshaw on the Moon.

Apollo 14 Landing Area—Fra Mauro Highlands
Touchdown—3:18 a.m. CST, Friday, Feb. 5, 1971
Coordinates—Latitude 3° 40′24″ S.
Longitude 17° 27′55″ W.
Lift-off from Moon—1:49 p.m. EST, February 6, 1971

Ed Mitchell sets his wristwatch in a very human moment before launch.

THE KANSAS CITY STAR

Walk on 'Rough' Moon

"There He Is"

Tiny Dip in Jobless Rate

'Perfect Descent'

Landing Is

Chicago Tribune

LONGEST WALK ON MOON

January Jobless Rate Dips

20 Schools Shut Down by Strike

2 Astronauts Seek Rocks, Stir Up Dust

Antares Sits on Loose Soil

TV Camera Shows Stroll

McCracken Backs ESP Predictions

Michigan Av. Shops Burn

"It certainly is a stark place here at Fra Mauro," Shepard said, as Antares rested in Apollo 13's original landing site just a handful of miles to the east of Apollo 12's site. It was dawn of the 14-day long lunar day when Antares touched down at the eastern edge of the Ocean of Storms. Shepard and Mitchell had a new gimmick with them to permit the collection of more soil and rock samples than "11" and "12" had gathered, a rickshaw kind of a pull-along cart. Until he was assigned to the Apollo 14 flight, Shepard had been grounded because of an inner-ear disorder that was later corrected by surgery. He had not been in space since his first suborbital flight in Mercury.

The Fra Mauro area was named for a 15th Century Italian Monk, a map-maker. Shepard and Mitchell made two moon walks—one of 4 hours and 48 minutes, the second of 4 hours and 35 minutes. Shepard made golf history by hitting a couple of golf shots on the moon with a club fashioned out of a piece of his lunar equipment. On the way home from the Moon, Mitchell conducted some ESP experiments with "receivers" back on Earth. After splashdown the crew underwent an isolation period to guard against Moon bugs, but none appeared, and no future Apollo Missions were put in quarantine.

Dave Scott at work with Hadley Rille and Hadley Delta in the background. St. George Crater is on the right in this view looking south.

APOLLO 15 "Endeavour and Falcon"

James B. Irwin *Pittsburgh, Pa.* *David R. Scott* *San Antonio, Texas* *Alfred M. Worden* *Jackson, Michigan*

Series: Apollo
Date: July 26-August 7, 1971
Crew & Age: CDR. Dave Scott (39) Gemini VIII, Apollo 9
CMP. Alfred M. Worden (39)
LMP. James B. Irwin (41)
Distinction: First landing in the lunar mountains • First use of the Lunar Rover • Worden launched subsatellite into lunar orbit • New Lunar Visit Record-66 hours, 54 minutes.
Crew Insignia: Apollo 15 Patch, U.S. Flag, NASA Insignia
Command Module: Endeavour
Lunar Module: Falcon

"The mission patch for Apollo 15 was basically designed by the Italian dress designer, Emilio Pucci. We had as a crew evaluated some 540 different designs for our crew patch. They appeared either too mechanical or too complicated or to have nothing to do with the flight, so finally, through a mutual friend, we asked Pucci if he would help us with the design. Now, Pucci, as I best recall, was an aeronautical engineer and had a good feeling for flight. With his artistic nature, we felt that he would be very helpful in the patch design. He did send us a design which was basically the same as the patch we eventually used, however, the colors were in the normal Pucci blues, purples and greens. We took his design, changed it from a square to a circular patch, made it red, white and blue and put a lunar background behind the three stylized birds that were the major Pucci contribution. The symbology is of three stylized birds flying over the lunar surface, each indicating one of us who were on the flight. The lunar surface behind the patch shows the landing site (next to Hadley Rille at the foot of the Apennine Mountains) and directly behind the stylized birds is a crater formation that spells "15" in Roman Numerals. You can also see from the stylized birds that they fly in formation with one on top and two closer to the lunar surface, indicating those who actually landed.

The Lunar Module was named for the mascot of the U.S. Air Force, the Falcon (this was an all Air Force crew).

The Command Module was much more difficult and we considered many different names trying to find a name which signified the exploration and at the same time was short, simple and had impact. A geologist, Farouk El-Baz, was working with us and one day he picked up a children's story book of major explorers and from that book came the story of Captain Cook and his scientific voyages to the Pacific on the ship, Endeavour. It seemed like a natural name and we chose it for the name of our Command Module."

—Al Worden
Command Module Pilot

In describing his voyage, Captain Cook said: "I had the ambition to not only go farther than man had gone before, but to go as far as it was possible to go."

Jim Irwin works at the Lunar Rover with Mount Hadley in the background.

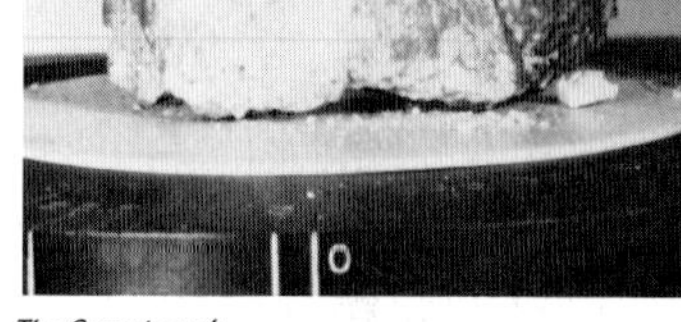

The Genesis rock.

Apollo 15 Landing Area—The Apennine Mountains, Marsh of Decay
Touchdown—6:15 p.m. EDT, July 30, 1971
Coordinates—26 degrees 6 minutes North Latitude
3 degrees, 39 minutes East Longitude
Lift-off from Moon—1:11 p.m. EDT, August 2, 1971 (Lunar Lift-off Televised)

INDIANA FINAL

The South Bend Tribune

THREE COSMONAUTS LAND—DEAD

McNamara Plea Told

Soyuz 11 In 'Correct' Descent

Links Saigon Aid To N. Viet Attack

Turks Plan To Halt

Tragedy struck once more in space as the three Soyuz 11 Cosmonauts died during reentry after nearly 24 days in space. Cause was attributed to an improperly sealed hatch that allowed all of the air to escape from the spacecraft. The crew was not wearing pressure suits. They were Commander Georgy Dobrovolsky, 43, Flight Engineer Vladislav Volkov, 35, and Test Engineer Viktor Patsayev, 37. It was the second straight Soviet space failure. Two months earlier the crew of Soyuz 10 had docked with Salyut 1, the unmanned space station, but were unable to enter due to a hatch malfunction.

The Miami Herald

Launch, Docking Go Smoothly

U.S. Drafting Bill to Halt Rail Strike

Apollo 15 Is Off to Explore Mountains, Rilles of Moon

Astronauts' Test Defeats Problem Of Faulty Switch

Medina's Trial Gets First Juror

Apollo's Schedule

America's 4th Manned Landing on the Moon left Earth at 9:34 a.m., Monday, July 26, 1971. It was the first of the "J" missions noted for "extensive scientific investigation of the Moon on both the Lunar Surface and from Lunar Orbit." Three-mile high mountains and a 1,200-foot-deep gorge almost as deep as the Grand Canyon marked the intended landing site at the edge of the Sea of Rains. The three earlier Apollo landings were all within 70 miles of the lunar equator, this voyage was 465 miles north at an area known as Hadley-Apennine. The Apennine Mountains where the astronauts landed are higher than our Sierra Nevadas or the Himalayan Mountains.

Al Worden conducted a wealth of lunar fact-gathering while Scott and Irwin were on the Moon. His instruments can be seen in the open portion of the Service Module.

Jim Irwin saluting America and its flag at Hadley Plain. Latitude 26°6'3" N., Longitude 3°39'10" E.

HOUSTON CHRONICLE

SUNDAY 35 cents

Irwin, Scott First to Ride on Moon

Scientists Praise Mobility

Massive Chicano Boycott Of HISD Schools Brewing

You've Heard Of Dog's Life? This Is It!

Deadline for Steel Strike Is Extended 24 Hours

Dave Scott and Jim Irwin landed at 6:15 p.m. Friday, July 30. "Okay, Houston," Dave radioed back, "The Falcon is on the Plain at Hadley." During their 67 hours on the Moon they drove the electric powered lunar rover 17½ miles on three geological field trips—6 hours, 33 minutes; 7 hours, 12 minutes and 4 hours, 50 minutes. On the 3rd EVA Scott gave the TV audience a practical demonstration of Galileo's discovery that falling objects, unhindered by atmospheric friction, drop at the same speed. The objects he used: his geological hammer and a falcon feather from the U.S. Air Force Academy's Mascot. Irwin accidentally stepped on the feather and the two then could not find it in the lunar dust. So even today, there is a falcon feather on the moon. We saw the lift-off from the Moon on the TV camera the crew left running when they left. On the return to Earth Al Worden did a spacewalk to retrieve film cassettes from the back of the Service Module that would be jettisoned upon reentry. Al was 197,000 miles from Earth during his spacewalk. A long first step, indeed. When the crew returned they proudly displayed the white anorthositic "Genesis Rock" they had found sitting conveniently up on a pedestal like formation on the moon as if waiting for their arrival. It is 4.1 Billion years old!

Veils: Mysterious Cover Ups Stage a Comebac

Chicago Tribune

HOME

Plan U. S.-Russ Space Tie

Docking Device Studied

Task Force Report

Race Av.: Tense Frontier Between Whites, Blacks

All the Comforts of Home

Luxury Comes to the Campground: 'Roughing It' Is a Lot Smoother Today

Two days after Apollo 15 splashed down (one of its 3 chutes failed on reentry), this hopeful headline greeted the World. The article said: "Joint space missions could mean the end of the Space Race started by the Russians when they launched the first satellite, Sputnik, in 1957 and won by America when Neil Armstrong and Edwin Aldrin landed on the moon in July, 1969." One of the first U.S.-U.S.S.R. compromises involved using colored running lights on the spacecraft as used on U.S. craft. The Soviets use blinking white lights. "...the Russians were surprisingly open in discussing their space program, including all of its problems...." The article concluded that it might mean the cancellation of Apollo 17 so the money (about $455 million) could be used on the joint flight. The cylindrical adapter was the key to this proposed mission, and was expected to take two years to design and build. The Apollo-Soyuz flight of July 15, 1975 was the result.

A magnificent view of America taken "outward bound" on Apollo 16. Note Lakes Michigan and Superior, the State of Florida and most of the southern and western United States. Spring was upon the land.

APOLLO 16 "Casper and Orion"

Thomas K. Mattingly II *Chicago, Illinois* — *John W. Young* *San Francisco, California* — *Charles M. Duke, Jr.* *Charlotte, North Carolina*

Series: Apollo

Date: April 16-27, 1972

Crew & Age: CDR. John Young (41) Gemini III, Gemini X, Apollo 10
CMP. Thomas K. Mattingly II (36)
LMP. Charles M. Duke (36)

Distinction: Highest Lunar Landing at Elevation of 25,688 feet • Set Lunar Speed Record—11.2 MPH • Set Lunar Distance Record—22.4 miles covered • 213 lbs. of Lunar Material was returned for study.

Crew Insignia: Apollo 16 Patch, U.S. Flag, NASA Insignia

Command Module: Casper

Lunar Module: Orion

"This patch was designed by Barbara Matelski in the graphics shop at Johnson Space Center. The idea came from us. We wanted to tell the teamwork story, plus identify the crew, plus keep America visible. The Eagle and shield and the red, white and blue were for the U.S.A. The wishbone for NASA. (This is the flight symbol from the NASA insignia). The moon to signify our landing. The 16 stars for our flight number. Ken selected Casper for his Command Module and John and I selected Orion. We wanted something connected with the stars."

—Charlie Duke
Lunar Module Pilot
10th Man on the Moon

Casper, the Friendly Ghost
© Harvey Publications, Inc.

Following a normal launch and a mysterious shredding of the Lem's Skin on the voyage out to the Moon, Apollo 16 had its share of unexpected glitches just like nearly every other spaceflight. After the lunar module Orion separated from T.K.'s friendly ghost, Casper, he had trouble returning to his 65 mile circular orbit from the 10 mile height he had dropped down to in order to release the lunar lander as low as possible to provide fuel for additional hover time. It took about 5½ hours to evaluate the problem, putting Charlie Duke and John Young behind time in their landing and subsequent exploration. They finally landed about 8:23 p.m. Thursday evening, April 20, 1972. They had fuel for 100 seconds of hover time left when they landed. The engineering of landing on the Moon is a finite business, indeed. It can also be humorous, albeit dangerous in retrospect. For Apollo 16 for the first time the Moon walkers had a beverage assembly inside their space helmets attached to a 32-ounce bag of potassium fortified orange juice. Charlie Duke's microphone got tangled in the orange juice tilt valve (straw-like device) and it released orange juice that drifted around inside his closed helmet in the weightlessness, coating his glasses. "The only thing bad about it," Charlie remarked, "is I got a helmet full of orange juice...." It was a sticky Moon landing for Charlie Duke.

"You may recall that the Teflon flight clothing was white and shapeless. I overheard some kids, who were watching one of the onboard T.V. shots of an earlier crew, remark that they looked like Casper, the friendly ghost. He was a comic character that has been around for a number of years. I liked the idea of something that was not so serious and which kids could identify with. So we named our Command Module "Casper."

—T.K. Mattingly II
Command Module Pilot

Charlie Duke collecting lunar samples on the rim of Plum Crater. The Lunar Rover is in the background.

CHICAGO Sun-Times

FINAL MARKETS

MOON ROMP

CLOUDY

Detroit Free Press

METRO

Action Line
Dial 222-6464

Apollo Explorers Blast Off After Record Trek on Moon

Rock Haul Excites Geologists

'Follies' Dominates The Tonys

Victor, Last of the Reuthers, Bids Sad Farewell to UAW

Duke and Young landed at the western edge of the Descartes Mountains about 30 miles west of the Kant Plateau. This is in the central lunar highlands, just a little south and to the right of our side of the Moon's face as we look at it. It is the highest spot on our side of the Moon. They were the happiest crew of Moon explorers since Al Bean and Pete Conrad on Apollo 12.

Young: "Look at those beautiful rocks...is this ever neat, Charlie."
Duke: "Yahoo, this is so great I can hardly believe it. We are proud to be Americans on an experience like this."
Young: "This is an indescribable experience. What's the difference between a hole and a crater?"
Mission Control: "Beats me."
Young: (Bumping into Duke): "By golly, we did it again. I never thought we'd run into each other on the Moon."

At 7:26 p.m. Sunday night Orion rocketed off the Moon, with people everywhere watching the departure on the TV attached to their electric car, Rover, parked a short distance away. Duke and Young took with them 245 pounds of rocks and lunar soil samples gathered on 3 Moon walks, 75 pounds more than Apollo 15 brought back. They had spent a record 71 hours and two minutes on the Moon's surface, four more hours than Apollo 15. Their first EVA was 7 hours, 11 minutes. Their second 7 hours and 23 minutes and the third 5 hours and 40 minutes. For T.K. Mattingly up above in Casper it was a satisfying trip after having gone through the crushing disappointment of being taken off the Apollo 13 crew at the last minute because of the measles scare. Most people did not remember that it was Charle Duke down on the Moon's surface on Apollo 16 who had exposed T.K. to the measles in the first place.

John Young does a "wheelie" during the "Grand Prix" workout of the Lunar Rover at a top speed of 6.2 miles per hour. At one point going down hill the Rover reached 10.5 MPH, a new Moon Land Record.

Apollo 16 Landing Area—The Cayley Plains, Descartes Region, Southern Highlands of the Moon.

Touchdown—8:23 p.m., April 20, 1972 CST
Coordinates—Latitude 8 degrees 59′ 29″ S.
Longitude 15 degrees 30′ 52″ E.
Lift-off from Moon—7:26 p.m. CST, April 23, 1972
(Lunar Lift-off televised).

John Young leaps high off the Moon to salute his flag and countrymen in an exuberant moment at Descartes. Latitude 8°59′29″ S., Longitude 15°30′52″ E.

Robert McCall tried a variety of ideas before he and the crew of Apollo 17 decided upon the final design of their patch. These are three of Bob's preliminary sketches. The Stonehenge theme was suggested by crew member Jack Schmidt. This 3,000 year old monument may have been an early astronomical calendar and foreteller of eclipses of the sun and moon. It is located on Salisbury Plain in Wiltshire, England. An early sketch of the final patch is also here. Note that the eagle and order of names were changed in the end and that several astronomical elements were added before the loom was turned on at the A-B Emblem Company in Weaverville, North Carolina.

APOLLO 17 "America and Challenger"

Dr. Harrison H. Schmitt *Santa Rita, New Mexico* — *Ronald E. Evans* *St. Francis, Kansas* — *Eugene A. Cernan* *Chicago, Illinois*

Series: Apollo

Date: December 7-19, 1972

Crew & Age: CDR. Gene Cernan (38) Gemini IX, Apollo 10
CMP. Ronald E. Evans (39)
LMP. Harrison H. Schmitt (37)

Distinction: Last Apollo Flight to the Moon • Record Time on Moon: 74 Hrs. 59½ Minutes • Command Module in record lunar orbit 147 hrs., 48 minutes • First Apollo Nightime Launch • EVA's totaling 22 hours and 4 minutes on Moon • Returned 250 lbs. of lunar material for analysis.

Crew Insignia: Apollo 17 Patch, U.S. Flag, NASA Insignia

Command Module: America

Lunar Module: Challenger

"The ideas for the Apollo XVII patch came six months before the final design. The crew had many philosophical ideas and ideas that we wanted to capture in a patch for the final Apollo moon mission. But it soon became evident that we did not have the artistic talent to put our words into pictures. Bob McCall, a personal friend of the crew's, offered to contribute his talent and time to help us in the development of this patch.

In this patch the crew wanted to capture Man, The Future, the Recollection of Apollo, the Flag and its contemporary heritage. The bust of Apollo represents the program itself; but in addition typifies man throughout his existence...his intelligence, his wisdom, and his ambition. Notice this bust is looking *forward.*

We have a contemporary American Eagle, the wings of which represent the flag of our country, the three stars representing the crew. The Eagle's wings just touch the lunar surface suggesting that this is a celestial body that man has visited and in the sense conquered. But it, too, is looking far beyond into the future; the future is represented by the Spiral Galaxy and Saturn implying that Man's goals in space will someday include the planets and perhaps, the stars. The Eagle is not resting on the laurels of its accomplishments of having gone to the moon, but rather is using what it has learned in leading mankind (represented by the bust of Apollo) into the future. Our main thesis was to capture man, our country, and the future. The colors of the emblem are red, white and blue, the colors of our flag; with the addition of gold, to symbolize the Golden Age of Space Flight that will begin with this Apollo 17 Lunar Landing. The Apollo image used was the famous Apollo of Belevedere sculpture in the Vatican Gallery in Rome.

The naming of the spacecraft was almost as complicated, but was arrived at more quickly. We wanted to find a way of paying tribute to the American public and also to the mission itself. We also wanted names with a tradition within the history of our country.

The Command Module was named "America". Besides great ships being named "America", we felt it was a way of identifying with and saying "Thank You" to the people of our country.

The name of the Lunar Module came down in the final analysis between the names "Heritage" and "Challenger." "Challenger" just seemed to describe more of what the future for America really held; and that was a challenge. The Lunar Module "Challenger" is appropriately resting on the mountain top overlooking the Valley of Taurus-Littrow, where it found its greatest glory."

—Gene Cernan

Commander, Apollo 17

12th Man to land on the Moon and the last to leave the lunar surface.

The last man on the Moon, Gene Cernan, places the Apollo 17 Mission Patch on his T-38 jet aircraft at Patrick Air Force Base south of Cape Canaveral a few days before launch.

Jack Schmitt next to the flag with Earth over his left shoulder, looking mighty far away.

Harry Truman's Condition Is Critical

THE BAY CITY TIMES

APOLLO HEADS FOR MOON AFTER DELAYED LAUNCH

Lunar Landing Still Expected to Be on Time

Astronauts Will Have Quiet Day

The last of our pioneering flights to the Moon began after a three hour delay due to a cranky computer. The night launch finally took place at 12:33 a.m. Ron Evans, the Command Module Pilot, was known to his friends as "Captain America" in honor of the Apollo, codenamed "America". Gene Cernan was making this third flight into space, his second back to the Moon. With Gene and Ron was rookie Astronaut Jack Schmitt, who had a Doctorate in Geology from Harvard University. The scientists had won out over the test pilots for this final flight and Schmitt had replaced X-15 pilot Joe Henry Engle who had originally been in training for the flight. The spectacular night launch was seen as far as 200 miles away in Miami where one observer said: "It looked like a Fourth of July firecracker shooting into the sky. It was a hell of a spectacular sight." Joe Engle had to wait another 10 years, but finally was the Commander of the second Space Shuttle flight.

ADDITIONS BE A SCHOOL VOLUNTEER 241-4651

Orlando Sentinel

FINAL Late Sports

'Tis a Privilege to Live in Central Florida

THE PROS

Miami 23	Detroit 21	St. Louis 24	N. England 17	San Fran. 20
N.Y. Giants 13	Buffalo 21	Los Angeles 14	N. Orleans 10	Atlanta 0
Chicago 21	Green Bay 23	Denver 38	Kansas City ... 24	Pittsburgh 9
Philadelphia ... 12	Minnesota 7	San Diego 13	Baltimore 10	Houston 3

6½ Minute Rocket Burst Puts Apollo In Moon Orbit

'Challenger' Attempts Landing Today

Truman Improves, Then Slips

Crew Eager To Begin Exploration

Demo Backer

Ron Evans has said: "Even though the space program was for all mankind, the good ole U.S.A. did it and I wanted my S/C to be named "America". Our flight was really not the end, but the beginning of a challenge to the future and so we named our LM "Challenger". As the trio swooped over the Moon's surface preparatory to the lunar landing Evans said they were just a little excited: "We're breathing so hard, the windows are fogging up on the inside." Schmidt interrupted a description of the lunar surface from orbit with "Hey, I just saw a flash on the lunar surface. It was right out there north of Grimaldi. It was a little bright flash near that crater right there at the edge of Grimaldi. It was just a pinprick of light. I was planning on looking for those kind of things." He asked the ground to check seismometers for the possibility that the flash was caused by a small impact, but the Moon was still vibrating from the impact of the Saturn 5 third stage and instruments left on the Moon by earlier Astronauts could not unmask the possible impact Schmidt saw. "Just my luck," he commented.

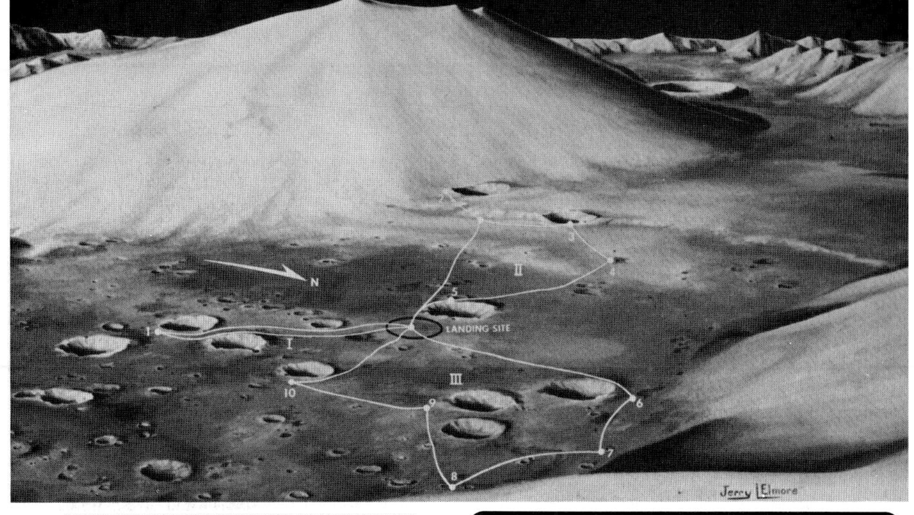

Al Shepard wishes Jack Schmitt well on the morning of launch.

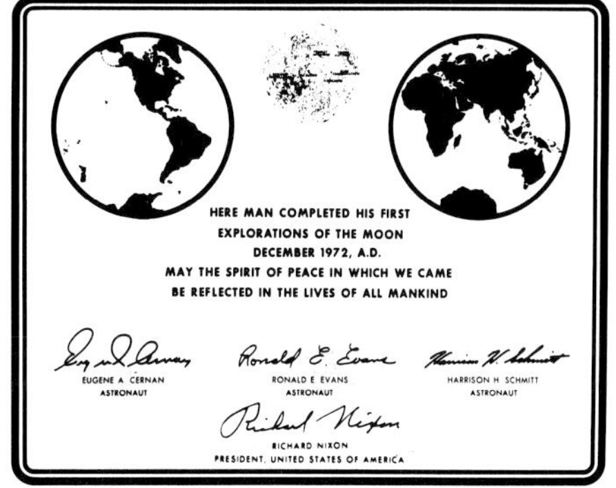

This plaque remains on the Moon with the descent stage of the lunar Module "Challenger" waiting for the men and women of Planet Earth to return.

As Gene Cernan finally stepped onto the Moon he said: "I think the next generation ought to accept this as a challenge. Let's see them leave footsteps like this." And after an exhausting 22 hours walking on the surface and after collecting over 250 pounds of rocks and soil, Chicagoan Cernan became the last man to leave the Moon with these words: "This is Gene and I'm on the surface. And as I take these last steps from the surface back home for some time to come, but, we believe not too long in the future, I'd like to just let what I believe history will record...that America's challenge of today has forged Man's destiny of tomorrow. And as we leave the Moon and Taurus-Littrow, we leave as we came, and, God willing, as we shall return. With peace and in hope for all Mankind. Godspeed the crew of Apollo 17." Enroute back to Earth, Ron Evans did a spacewalk to retrieve 3 film cans from the rear of the service module. "Hot diggety-dog! Evans shouted." Am I on the tube? Did you see me wave? Hello, Mom...Hello, Jon, Hello, Jaime," Ron called out to his wife, 11-year-old son and 13-year-old daughter. The cannisters held 1.6 miles of film, included a cross-section radar view of the top mile of the Moon's crust. The Command Module, America, splashed down into a calm sea at 1:25 p.m. CST December 19, 1972 successfully completing the United States First Moon Landing Program. It was a nice Christmas present. Enroute back from the Moon, Gene Cernan had said: "It (Apollo) has been a beginning. I don't think there ever will be an end, not as long as man is alive and willing."

COLD

Detroit Free Press

STATE

EVANS TAKES SPACEWALK

Action Line

Apollo to Earth: 'Hello, Mom'

Pilot Is Finally In Limelight

Truman's Kidneys

Crash Kills Two Swimmers

SMEAT

A Skylab Medical Experiments Altitude Test was held for 56 days beginning July 26, 1972 to obtain baseline medical data and evaluate medical experiment equipment planned for use in the upcoming Skylab program. Astronauts Robert L. Crippen, William E. Thornton and Karol J. Bobko participated in the test in the 20-foot altitude chamber at the Manned Spacecraft Center in Houston. The test ended September 20.

Ken Snyder, a SMEAT test conductor, played a "move a day" chess match with Astronaut Bob Crippen inside the altitude chamber. Each day's move was passed through to the crew.

The SMEAT crew entering the Skylab simulator at the beginning of their 56 day "flight."

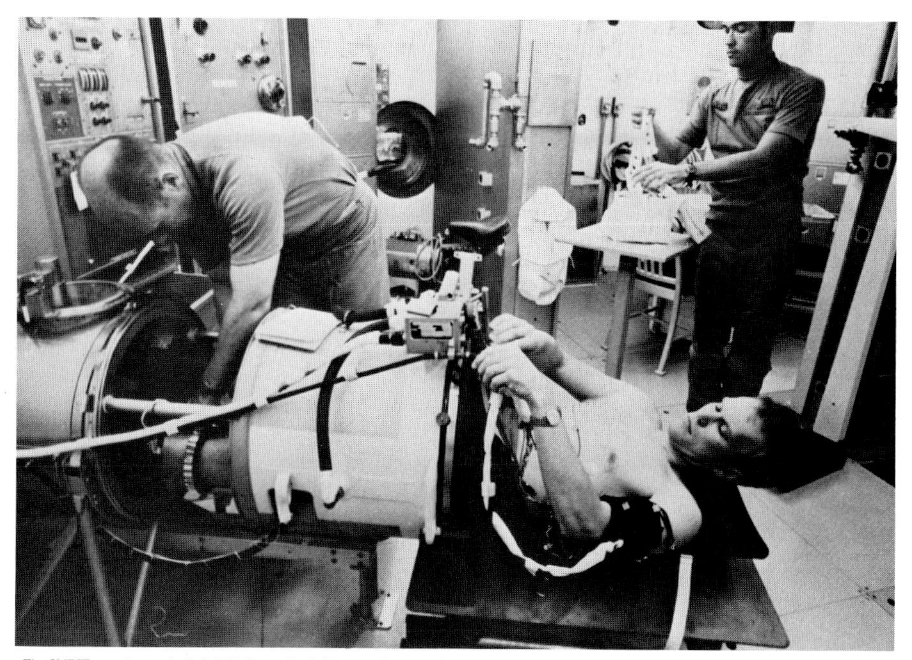

The SMEAT experiments included this Lower Body Negative Pressure Test with Astronaut Karol Bobko as the subject.

The Skylab Program

Skylab was the name of America's first orbiting space station. It used hardware developed for the Apollo moon landing including the command module and the Saturn V rocket.

Three crews sailed on Skylab on the longest space voyages up to that time lasting 28, 59 and 84 days each. The name Skylab was proposed by Donald L. Steelman, U.S. Air Force, while he was assigned to NASA. And indeed, it was just that, a laboratory in the sky with astronaut-scientists aboard, as well as astronaut-pilots.

The Skylab space station remained in orbit until July 11, 1979 when it finally fell to earth in southwestern Australia.

An artist's view showing how the Apollo's docked at the end of the Skylab Space Station.

Skylab Hardware

Our Skylab Space Station was composed of five modules:

1. The Orbital Workshop—This was a coverted upper stage of the Saturn V rocket and contained all of the comforts of home including work areas, a kitchen, sleeping quarters, bathroom and even a shower. It was the size of a small 3 bedroom house and was 48 feet long and 21½ feet in diameter.
2. The Airlock Module—This was 10 feet in diameter and 17½ feet long and contained electrical power and communications systems and the airlock through which the astronauts exited to perform duties outside Skylab during the mission.
3. The Apollo Telescope Mount—Contained the 8 solar telescopes, the control gyros and four solar array wings. The crew controlled and monitored these solar telescopes from within the Multiple Docking Adapter.
4. Multiple Docking Adapter—This is what the Command Module docked to, it contained two docking ports, an extra in case of an emergency rescue mission. It also had the ATM control console and controls and sensors for Earth resources viewing.
5. Command and Service Module—This was basically the same Apollo ship as used on the Moon flights and had been modified to ferry supplies up to and down from the Skylab. The Service Module was jettisoned just before reentry and burned in the atmosphere. A special rescue configuration is shown here that never had to be used. This would have permitted a rescue team of two astronauts to rocket up to meet and return with the three stranded crew members in case their own Apollo failed. This was almost done during the second Skylab Mission, but a problem with their Apollo was solved before this became necessary.

This double exposure shows the Skylab atop its Saturn on the right, and the Skylab Apollo on its shortened version of the rocket and "milkstool" launch platform on the left. The Command Module by itself did not require the huge rocket of its Moon Mission days.

The Workshop and Crew's Quarters portion of Skylab is shown here inside the Vehicle Assembly Building at the Cape prior to mating to its Saturn rocket.

SKYLAB RESCUE CSM GENERAL ARRANGEMENT

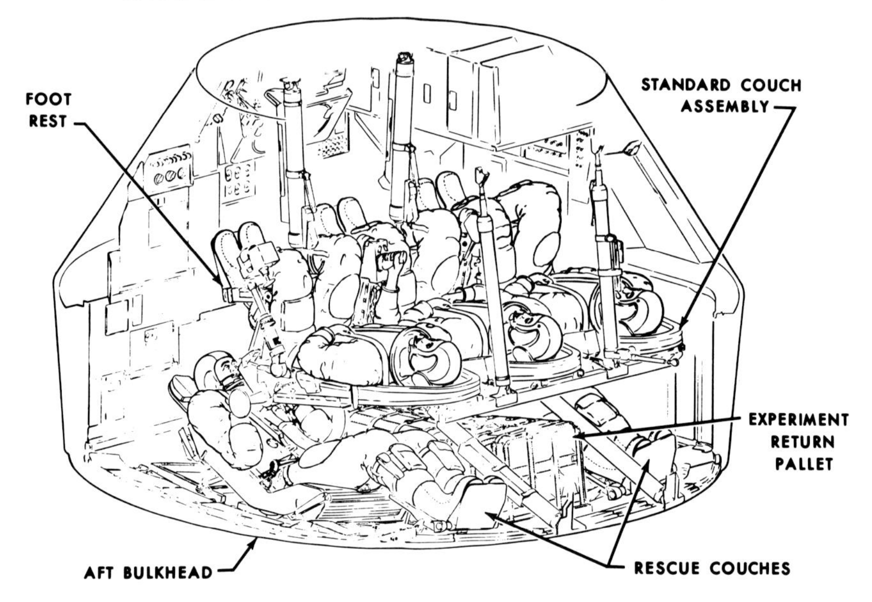

The rescue configuration of the Apollo that almost had to be used on the flight of the second Skylab crew.

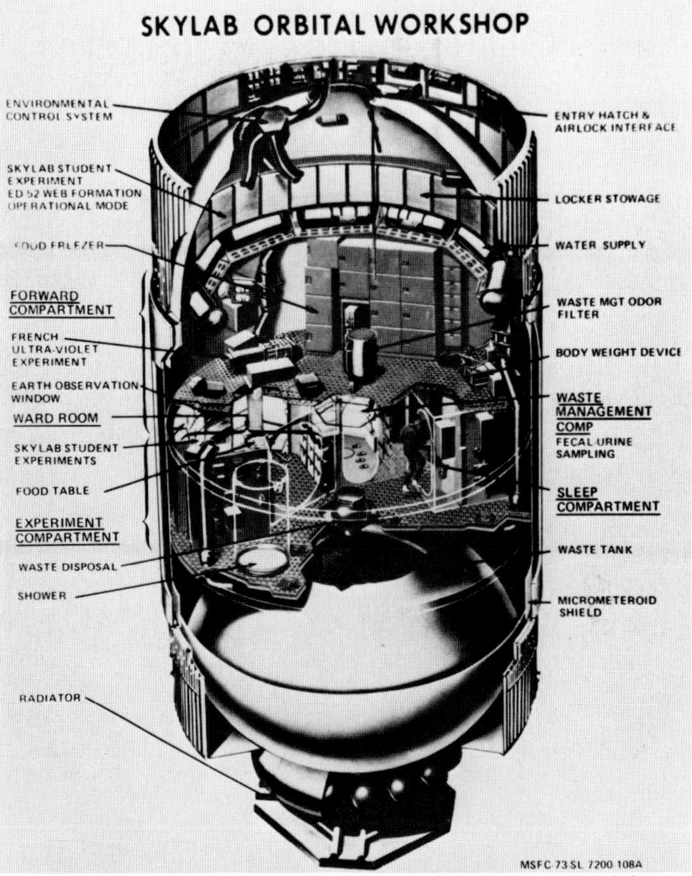

A cutaway view of the interior of the workshop and Crew's Quarters of the Skylab. The huge tank on the bottom served as the garbage can.

Grayling, Michigan and surrounding area from 270 miles overhead. Taken by the first Skylab crew. Portions of Lakes Michigan, Huron and Superior can be seen. The author lived in Grayling when he began this book.

In 1923 Hermann Oberth suggested the idea of space stations and on May 14, 1973 Skylab I, an unmanned Orbital Workshop was launched from Pad 39A using the last Saturn V to be flown. 63 seconds into the flight the meteoroid shield tore loose and jammed one of the workshop's solar wings. It continued on up, to orbit above the Earth every 93 minutes at 270 miles high. From here it would fly 50 degrees above and below Earth's equator, approximately from Montreal, Canada down to the southern tip of South America. It would fly over 75% of the Earth's surface and about 90% of its population. Because of the incident at launch, the first crew of Astronauts did not rocket up to inhabit Skylab the next day as planned. They first had to decide how to fix the critter, or Skylab could have been a very expensive empty tin can.

CLOUDY

Detroit Free Press

STATE

Action Line Dial 222-6464

Astronauts Put on Masks, Begin Sniffing Out Skylab

Clean Air Found in First Room

Lindbergh and His Wife:

Finally, after a great deal of study and practice on the ground, the first crew of Astronauts rocketed up to track down Skylab on May 25. Commander Pete Conrad radioed back, "Tallyho, the Skylab. We got her in daylight at 1.5 miles, 29 feet per second." It had taken them 8 hours of maneuvering to close. They docked with the monster station, its largest section, the workshop, was 48 feet long and 21½ feet in diameter. The "cluster of cans" as some referred to the complete station was as big as a 3 bedroom house, weighed 85 tons and was as tall as a 12 story building. But because it was unprotected from the heat of the sun in some areas because of the mishap there was serious concern that deadly gases had formed inside. The crew docked and spent their first night asleep in the Command Module before attempting to enter the cavernous space station. When they did, they put on gas masks and found the temperature inside one area to be around 130 degrees, but there were no toxic gases. Using a scientific airlock on the side of the station, they erected a parasol thermal shield (an umbrella) to keep the searing heat of the sun from the damaged Skylab. The day before back on Earth, the Soviet newspaper Pravda announced that there would be a joint American-Russian flight on July 15, 1975. Pravda said both countries had agreed on radio frequencies to be used.

SKYLAB 1 Officially Skylab 2

Dr. Joseph P. Kerwin
Oak Park, Illinois

Charles Conrad, Jr.
Philadelphia, Pa.

Paul J. Weitz
Erie, Pa.

Series: Skylab

Date: May 25-June 22, 1973

Crew & Age: CDR. Pete Conrad (42) Gemini V, Gemini XI, Apollo 12*
Science Pilot Joseph P. Kerwin (41)
Pilot Paul J. Weitz (40)

Distinction: Set new world record 672 hrs. 49 min. in space • Deployed "parasol" thermal shade to save Skylab • Deployed the jammed solar wing • Kerwin became first Medical Doctor in Space

Crew Insignia: Skylab 1 patch, U.S. Flag, NASA Insignia

Command Module: Apollo

Space Station: Skylab

"Our patch was executed by science fiction illustrator Kelly Freas. He wrote his story of the design in the June 1973 issue of "Analog." We did not name our vehicles. We figured CSM's were only for up and down, so none were required. Our Skylab cluster was used for 3 crews and we made one half-hearted attempt at unanimity for one name from 9 guys, but no one was over enthusiastic so the effort died."

—Paul Weitz
Pilot

"The kinds of ideas we tossed around emphasized that it was to be a peaceful mission; that in addition to doing "pure" science a la Apollo, we're doing work that will directly benefit Mother Earth and its citizens; and that, in a very real sense, we are doing more than just exploring near—Earth space—we're homesteading it, preparing to live and work up there. Thus, droves of people, Earth scenes, optical devices, covered wagons, plows and log cabins have all come to mind."

An excerpt from *Analog* magazine, June 1973, by Kelly Freas. —Dr. Joe Kerwin
Science Pilot

The patch shows the Skylab silhouetted against the Earth's globe, which in turn is eclipsing the Sun—showing the brilliant signet-ring pattern of the instant before total eclipse.

Officially the Skylab manned missions are still listed as Skylab 2, 3 and 4 with Skylab 1 being the initial launch of the Skylab Space Station itself. However, to avoid public confusion the patches for the manned Skylab missions are numbered 1, 2 and 3.

*On June 2nd, Pete Conrad celebrated his 43rd birthday while orbiting in the Skylab space station.

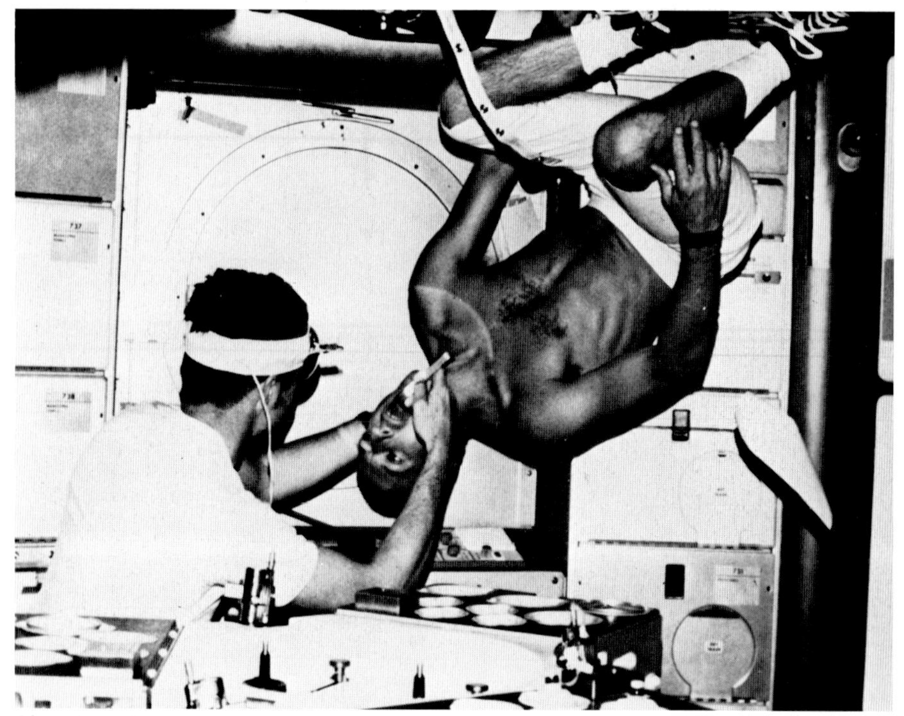

A far-out "house call"—Dr. Kerwin gives Pete Conrad a medical exam during the Skylab I Mission. Note the floating piece of paper below Pete.

WARM

Detroit Free Press

STATE

ASTRONAUTS FREE WING

Risky Space Repair Works

Action Line Dial 222-6464

Nixon Bug Data Second Hand, Haldeman Says

On June 7, two of the crew, Conrad and Kerwin, donned their spacesuits and went outside Skylab. They assembled a 25 foot long aluminum pole from 5 sections and attached a cable cutter tool to one end. After trying for quite some time to cut some aluminum debris on the solar-wing beam it suddenly let go propelling Conrad into space. His safety tether saved him. A few minutes later, as the two struggled with the solar beam the clevis bracket on the actuator broke, freeing the beam and tumbling both Conrad and Kerwin into space. Again, the safety lines held them. Within 6 hours the power capacity had been increased from 4,000 to 7,000 watts in the solar wings, saving the Mission.

The Detroit News

Ostracized West Pointer still graduates

Cadet beats 'ordeal of silence'

Skylab saved from freezing by astronauts

Haldeman differs with President

After their hard day's work the crew was sound asleep when they received an Emergency Call from Mission Control that they were in danger of their pipes freezing. So like tourists at a lake cottage they improvised by heating their cooling loop with their special EVA underwear draped over a hot water tank. The "liquid cooled garments" have small rubber tubes running through them and by heating them against the hot water tank they could in turn heat up the freezing pipes they were wrapped around. The problem had occurred because electrical equipment that would normally be used to keep the water warm had been turned-off to conserve electric power while Conrad and Kerwin had worked outside the space station the day before. Valves and regulators stuck closed. "Uncle Wiggly's airship" as author Henry S.F. Cooper, Jr. had described the crippled pin-wheel space station "sailed on." Crew #1 spent 28 days in scientific research in Skylab before splashing down 800 miles west of San Diego. Their daring EVA repair had saved the $2.6 billion project.

Skylab, showing repairs made to it by its first tenants—Conrad, Kerwin and Weitz. The "house in space" was damaged during launch and lost one of its solar wings, causing it to overheat. This sunshade was the solution.

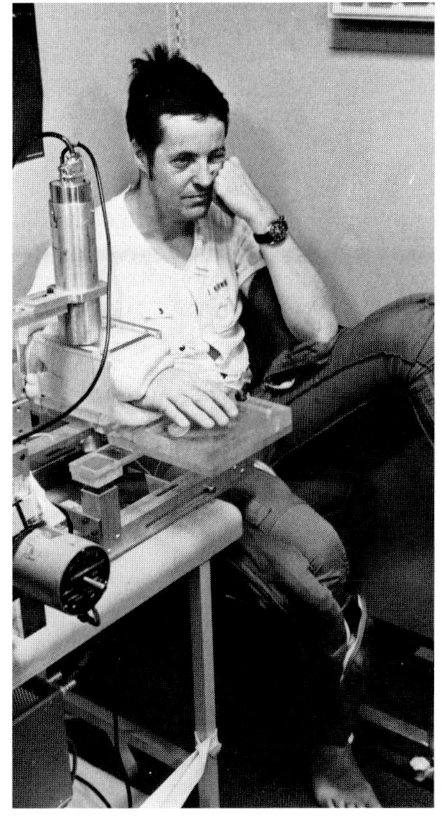

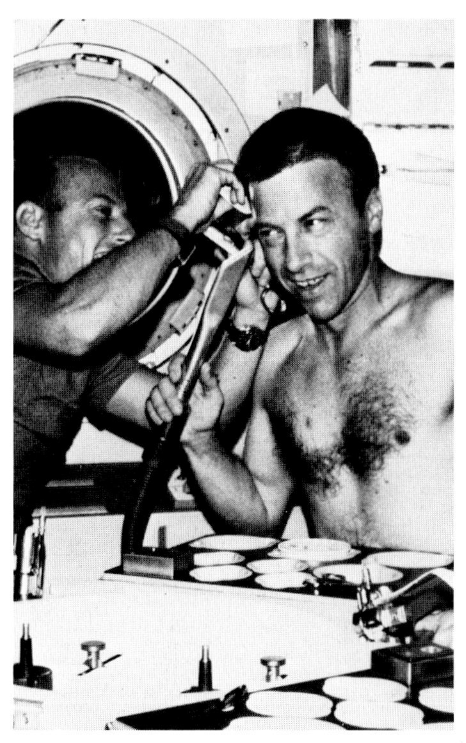

Pete Conrad gives Paul Weitz a haircut during the Mission. Paul is holding the vacuum nozzle to capture the clippings in the weightlessness.

An exhausted Joe Kerwin undergoing a medical examination after splashdown.

DaVinci's man illustrates the proportions of the human form and suggests the many studies of man himself conducted on this mission in the zero-gravity environment of space. The two hemispheres represent the two additional areas of research—studies of the sun and the development of techniques for survey of the Earth's resources to be used in forestry, geology, hydrology, agriculture and other disciplines.

The work of Leonardo Da Vinci (1452-1519) seems particularly appropriate as a symbol of the Skylab II flight. He was not only a great artist, but also a farsighted naturalist, scientist and engineer. His studies of human anatomy were the most complete and accurate work in existence for several centuries. His talent was also directed to many of the practical problems of his country, including designs for hydraulic systems, canals and locks for commerce.

In aerodynamics, his study anticipated Newton's Third Law by about 200 years; and his ideas were incorporated in sketches of flying machines, based upon careful and extensive observations of birds in flight.

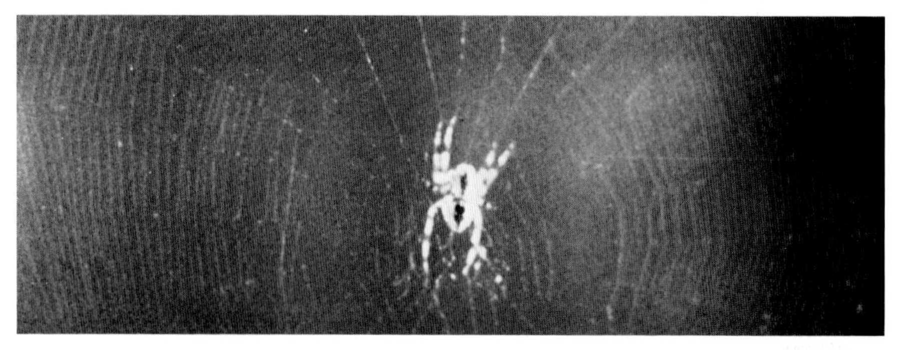

SKYLAB 2 Officially Skylab 3

Dr. Owen K. Garriott
Enid, Oklahoma

Jack R. Lousma
Grand Rapids, Michigan

Alan L. Bean
Wheeler, Texas

Series: Skylab
Date: July 28-September 25, 1973
Crew & Age: CDR. Al Bean (41)
Science Pilot Dr. Owen K. Garriott (42)
Pilot Jack R. Lousma (37)
Distinction: 59 days, 11 hrs. 9 minutes in space • Completed 3 two man EVA's • Invested 1,081 experiment hours

Crew Insignia: Skylab 2 Patch, U.S. Flag, NASA Insignia
Command Module: Apollo
Space Station: Skylab

"Initially we asked the art departments at Rockwell (LA), McDonnell Douglas at Huntington Beach (builders of the Skylab), Kennedy Space Center and others to search the windmills of their minds for our mission patch considering the major objectives of Skylab-Earth, Sun and Medical. The group at Huntington Beach was selected, although we received many good ideas from all of them.

We had decided that our patch should be red, white and blue for obvious reasons. The Earth half of the patch is pretty straightforward. The sun half is a little special in that the solar flare depicted in yellow orange is the shape of one Owen Garriott had done extensive analysis on years before. Leonardo Da Vinci's man represents the medical aspects of the flight. Certain modifications were made in DaVinci's art to make it more suitable for family viewing.

An interesting sidelight involves the wives' patch, a "first" which was done without our knowledge. The first time we saw the "wives'" patch was when we arrived in orbit and began to open the storage lockers in the Command Module to get our gear out. Neatly pasted to the interior of three of these locker doors were decals of the "wives" patch. This was a great idea and consistent with one of our mottoes, "Never lose your sense of humor."

—Jack Lousma
Pilot-Skylab 2

Arabella and her web. Taken in flight during week #3. High School student Judith Miles of Lexington, Mass. suggested this experiment as part of a National Student Program. 19 student experiments were flown onboard Skylab.

This is the lighthearted patch that flew aboard Skylab 2 designed to honor the Astronauts' wives. It carries the first names of Sue Bean, Helen-Mary Garriott and Gratia Lousma. A female spoof on DaVinci's universal man, as depicted on the crew patch. It was conceived by French journalist-artist Jacques Tiziou of Merritt Island and designed by Houston artist, Ardis Shanks Settle.

The Detroit News

Astronauts stay calm; rescue work is speeded

Floods kill 6 in N.J. downpour

From Tourist Council chief to consultant

State rehires forced-out official

The second crew to visit Skylab was headed up by Al Bean who had walked on the Moon during Apollo 12, by Owen Garriott who had taught electrical engineering at Stanford University and by Jack Lousma, a former Marine Pilot. They roared into the Florida sky at 7:00 a.m. July 28, 1973 and eight hours later Lousma called out, "Here's our home in the sky." With them went 6 pocket mice, 2 Mummichog minnows, 50 minnow eggs, 720 fruitfly pupae and two common Cross spiders named Arabella and Anita. They also carried a "six pack" of rate gyroscopes to repair Skylab's and a new umbrella. Trouble in the Apollo thrusters threatened an emergency rescue Mission. Two Astronauts would rocket up in a modified Apollo and ferry the three Astronauts already aloft back to Earth. But this did not prove necessary and the Mission continued for its full 59½ days and fulfilled 150% of its mission goals. The crew made 858 revolutions and covered 24,400,000 miles. They brought back with them 16,000 photographs and 18 miles of Earth resources tape. They also performed 333 medical experiments, after spending the first 3 days very ill with motion sickness.

On August 6, Owen Garriott and Jack Lousma erected a new solar shield over the parasol that Conrad and Kerwin had put up. They spent 6 hours and 31 minutes outside on this EVA, a new world record. NASA was afraid the first umbrella would deteriorate and not do its job. The crew also spent over 300 hours conducting astronomical observations and saw 100 solar flares. Arabella and Anita, part of a high school student's science experiment, both spun webs in zero-gravity. The crew pulled everyone's leg when they played back a tape that Garriott's wife had made before the launch pretending that she had stowed-away on board. The soft female voice from space had gotten Mission Control's attention for sure. The Skylab II crew landed southwest of San Diego on September 25. Total flight time was 1,427 hours 9 minutes and 4 seconds.

Three old-fashioned girls greet their men upon their return from a 59 day spaceflight at Ellington Air Force Base south of Houston and north of the Manned Spacecraft Center. Note the "Wive's Patch" on the rear of the car.

A unique variation of a Mission Patch. Al Bean's wife, Sue, and daughter, Amy Sue, 10, proudly display the Family Patch. Son, Clay, 17, joins them.

New Cracks Delay Skylab Launch

The San Diego Union FINAL

HOUSE OKs PIPELINE;
SENATE TO ACT TODAY

Police Jobs Flooded By Applicants

Kissinger, Mao Hold Long Talk

Move To Reduce Regulatory Powers Rejected

The 3rd Skylab Crew missed their launch date by about a week when cracks were discovered in their Saturn 1-B rocket as this headline points out. Finally, on the morning of November 16, 1973, with 160 pounds of groceries safely stored aboard, the trio lifted off the flat Florida beachside. All three crewmen were rookies. But before they were done they would spend 12 full weeks in space. It was America's 30th Manned Spaceflight. Both Pogue and Carr had motion sickness the first 3 days as had the previous crew. On Thanksgiving Day, Gibson and Pogue did a 6½ hour EVA repairing an antenna and in replacing film in the solar observatory. On December 13 they spotted the Comet Kohoutek and Ed Gibson described it as "one of the most beautiful sights in creation I've ever seen." Meanwhile, in Washington they were talking about building a pipeline from the North Slope of Alaska to Valdez.

When Skylab's last crew splashed down on February 8, 1974 they had completed the longest space exploration to date—84 days 1 hour and 15 minutes during 1,214 orbits of Earth. Their distance traveled was 34.5 million miles. They completed 56 experiments, 26 science demonstrations and 13 student experiments during their flight. All in all the three Skylab crews took 40,286 Earth resources photos and 182,842 solar observatory photos. Skylab supported its crews for 171 days of space travel. A record at the time. Ed Gibson sort of summed things up during the Mission: "Being up here and being able to see the stars and look back at the Earth and see your own Sun as a star makes you realize the universe is quite big, and just the number of possible combinations which can create life enters your mind and makes it seem much more likely."

SKYLAB 3 Officially Skylab 4

Gerald P. Carr
Denver, Colorado

Dr. Edward G. Gibson
Buffalo, New York

William R. Pogue
Okemah, Oklahoma

Series: Skylab
Date: November 16, 1973-February 8, 1974
Crew & Age: CDR. Gerald P. Carr (41)
Science Pilot Dr. Edward G. Gibson (37)
Pilot William R. Pogue (43)*
Distinction: Set Record for Most Traveled Humans of all time with 34,469,696 miles, also longest time spent weightless at 84 days, 1 hour, 15 minutes, this Proved Man Can Medically Go To Mars.
Crew Insignia: Skylab 3 patch, U.S. Flag, NASA Insignia
Command Module: Apollo
Space Station: Skylab

"The three of us dreamed up the concept. Bill Pogue put together the description, and Barbara Matelski from the Johnson Spacecraft Center Graphics Department did the artwork. We gave our concept to several artists, but none of them came up with a design that delivered the message, so we rough sketched our idea, and Barbara put it together."
—Jerry Carr
Commander, Skylab 3

"The symbols in the patch refer to the three major areas of investigation in the mission. The tree represents man's natural environment and relates to the objective of advancing the study of earth resources. The hydrogen atom, as the basic building block of the universe, represents man's exploration of the physical world, his application of knowledge, and his development of technology. Since the sun is composed primarily of hydrogen, the hydrogen symbol also refers to the Solar Physics mission objectives.

The human silhouette represents mankind and the human capacity to direct technology with a wisdom tempered by regard for his natural environment. It also relates to the Skylab medical studies of man himself.

The rainbow, adopted from the Biblical story of the Flood, symbolizes the promise that is offered to man. It embraces man and extends to the tree and the hydrogen atom, emphasizing man's pivotal role in the conciliation of technology with nature by a humanistic application of our scientific knowledge." *From the Skylab III Astronaut's Calling Card*

*Bill Pogue observed his 44th birthday on January 23, 1974 while orbiting the Earth aboard the Space Station, Skylab.

A bearded space station Commander, Jerry Carr, and Ed Gibson perform a magic trick in the zero-gravity Skylab.

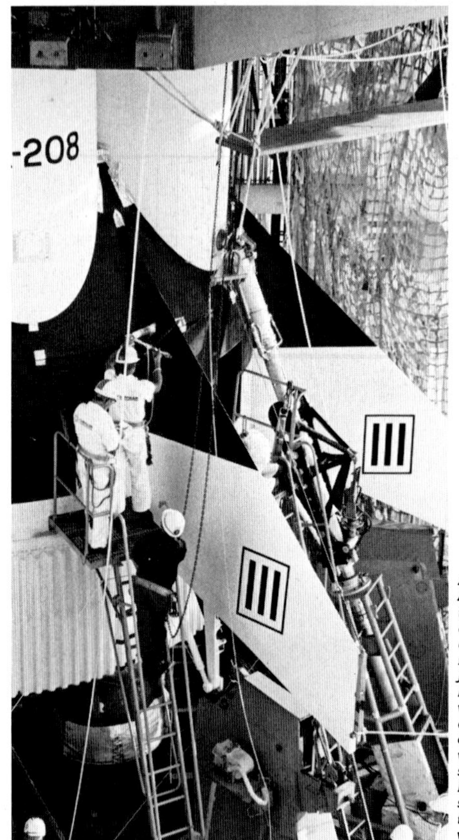

Fire Marshals Cite Four UF Buildings — Page 5A

Final Edition

Gainesville Sun

Celestial Fireworks at Skylab's Death

Carter Appears To Be Wrapping Up Policy Talks

Skylab 'Fell All Around Us'

At a little after noon E.D.T. on Wednesday, July 11, 1979 (after midnight Australia time) a flaming Skylab completed its Mission when it reentered the Earth's atmosphere and disintegrated in its death plunge, spraying debris over the most desolate area of Australia. The space station had completed nearly 35,000 orbits during its 2,249 days as a man-made moon, traveling around 1 billion miles. "It was an incredible sight," said Jack Seiler, a rancher in Australia's vast outback. "Hundreds of shining lights dropping all around the homestead... we could hear the noise of wind in the air as bigger pieces passed over us. Just after the last pieces dropped out of sight, the whole house shook three times... the horses on the property ran mad. They galloped all over the place and the dogs were barking." Seiler's Noondonnia Station, a 200 square mile cattle and sheep ranch is about 10 miles north of Balladonia, 530 miles east of Perth. Another observer, Bradley Smith of Perth's Bickley Observatory said that from where he watched it looked twice as large as the Moon. "A train on fire with bits of fire burning all the way down the carriages, that's what it was like," Smith said. "There were golds, yellow, reds... we were damn lucky to see it."

Commander Jerry Carr and his crew patch on the transfer van.

Man's last close look at Skylab. This photo was taken by the last crew as they left to return to Earth. The circular area is the docking port.

Seeds of the Future

The American and Soviet spacemen exchanged tree seeds while in orbit. The American seeds were White Spruce gathered by Hans Nienstaldt of the Forestry Science Lab in Rhinelander, Wisconsin. They were the first generation of a genetically improved selection. They have been planted in Moscow in the Cosmonaut Park in honor of the Apollo-Soyuz flight.

APOLLO-SOYUZ

First International Space Rendezvous

The ASTP crews: left to right, sitting: Donald K. "Deke" Slayton, Sparta, Wisconsin; Vance D. Brand, Longmont, Colorado; Valeriy N. Kubasov, Vyazniki, Soviet Union. L to R, standing: Thomas P. Stafford, Weatherford, Oklahoma, and Aleksey A. Leonov, Listvayanka, Alta Kray, Soviet Union. Leonov was the first man to walk in space in March, 1965.

Series: Apollo-Soyuz Test Project
Date: July 15-July 24, 1975
United States
Crew & Age: CDR. Tom Stafford (44) Gemini VI, Gemini IX, Apollo 10
Docking Module Pilot Deke Slayton (50)
CMP. Vance Brand (44)
Soviet Union
Crew & Age: Cmdr. Aleksey Leonov (41) Voskhod 2
Flight Engineer Valeriy Kubasov (40) Soyuz 6

Distinction: 1st International docking in space • 1st coordinated launch of two spacecraft from different countries
Crew Insignia: ASTP Crew Patch
U.S. Command Module: Apollo
Soviet Command Module: Soyuz

"The ASTP (Apollo-Soyuz Test Project) insignia was designed by the Soviets. The emblem has the words Apollo in English and Soyuz in Russian around a center disc which depicts the two spacecraft docked together in Earth orbit. The Russian word "Soyuz" means "Union" in English. (This is the original Soviet art rendering).

Our ASTP crew patch (shown above) was a composite of Rockwell International and NASA ideas. Stan Jacobsen at the Johnson Spacecraft Center provided us with the final drawing of the patch. The patch carries the names of the three American and two Soviet crewmen and the words Apollo in English and Soyuz in Russian around an artist's concept of the Apollo and Soyuz spacecraft about to dock to Earth orbit. The white stars on the blue background represent the American Astronauts, the gold stars on the red background represent the Soviet Cosmonauts.

Tom Stafford, Deke Slayton and I presented the concept of the patch to the Soviet crew during a pre-mission training session. Leonov and Kubasov agreed that the design should become the crew patch for the International Mission."

—Vance Brand
Command Module Pilot

During a short stop on a ride from Star City to Moscow during training, the American and Russian teams engaged in a friendly snowball fight. Star City is 40 minutes by car northeast of Moscow and is the Cosmonaut training center.

Reprinted with permission of Parade Publications, Inc.

Detroit Free Press

parade

How Much Is an Oscar Worth?

State Laws That Create Higher Prices

U.S. and Soviet Crews Eagerly Await Joint Space Flight

April 6, 1975 3:20 P.M. Moscow Time (8:20 A.M. EDT), Tuesday July 15, 1975

"This is the Soviet Mission Control Center. Moscow time is 15 hours, 15 minutes. Everything is ready at the Cosmodrome for the launch of the Soviet spacecraft Soyuz. Five minutes remaining for launch. Onboard systems are now under onboard control. The right control board... opposite the commander's couch is now turned on. The cosmonauts have strapped themselves in and reported that they are ready. They have lowered their face plates. The key for launch has been inserted... the crew is ready for launch." (five minutes later)... "Ignition. The engines are powered up. The launch... the booster is off. Moscow time 15 hours, 20 minutes, 10 seconds. The flight is proceeding normally."

The launch of Soyuz 19-Baykonur Launch Complex; Moscow Mission Control Center-Kaliningrad.

THE BAY CITY TIMES

Thumb Cattle Quarantined

St. Helen Shooting of Man by Deputies Remains Unexplained

American, Russian Craft Rendezvous in Space

It's News Today

Monkeys Kidding Around

Apollo lifted off on-time for the historic rendezvous of an American and a Russian spaceship with many around the World hoping that one day the two adversary nations might cooperate together on a joint mission to Mars. Deke Slayton, one of the Original 7 Astronauts who was finally making his first spaceflight now that his heart murmur had disappeared said: "Man, I tell you, this is worth waiting 16 years for." And as Apollo 21 (there had been 3 Skylab Missions since Apollo 17) went into orbit, Vance Brand aboard the United States Spaceship exclaimed in Russian to the waiting Soyuz: "Mîy nakhoditsya na orbite!" (We are in orbit). The docking would occur on the 36th Soviet revolution after 29 American revolutions in the second day of flight. Speaking in Russian, Deke Slayton radioed to the Soviets: "Soyuz, Apollo. How do you read me?" Kubasov answered in English, "Very well. Hello everybody."

Slayton:	*"Hello, Valeriy. How are you. Good day, Valeriy."*
Kubasov:	*"How are you? Good day."*
Slayton:	*"Excellent... I'm very happy. Good morning."*
Leonov:	*"Apollo, Soyuz. How do you read me?"*
Slayton:	*"Alexey, I hear you excellently. How do you read me?"*
Leonov:	*"I read you loud and clear."*
Slayton:	*"Good."*

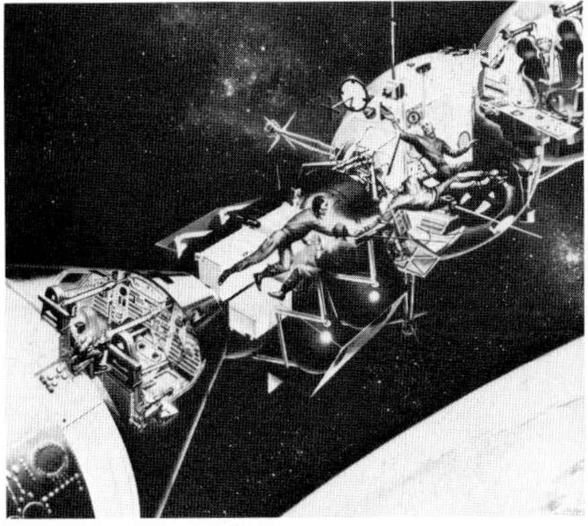

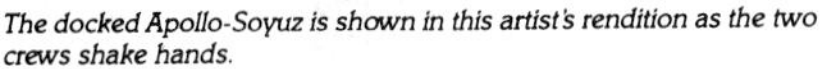

The docked Apollo-Soyuz is shown in this artist's rendition as the two crews shake hands.

Leonov breaks bread with the Americans during a visit to Apollo high above Earth. The crews took turns visiting one another's spaceships during the historic link-up.

The Soviet and American crews visited President Ford in the White House before the flight. Left to right: General Vladimir Aleksandrovich Shatalov, Chief, Cosmonaut Training; Cosmonauts Kubasov and Leonov; Soviet Ambassador Anatoliy Dobrynin; President Gerald Ford; Dr. George M. Low, NASA Deputy Administrator; and Astronauts Stafford, Slayton and Brand. September 7, 1974.

At 2:17 P.M. on July 17 Tom Stafford opened the hatch leading into the Soyuz and high above the French city of Metz the two commanders, Stafford and Leonov shook hands. In the background inside the Russian spacecraft there was a hand lettered sign in English—"Welcome aboard Soyuz." Later, in a message radioed-up to the crews from Leonid Ilyich Brezhnev, the Soviet leader, he said: "...the Soyuz Apollo is a forerunner of future international orbital stations." The American President, Gerald Ford, said that it had "taken us many years to open this door to useful cooperation in space between our two countries." And in talking to the crewmen individually, President Ford asked Deke Slayton: "As the world's oldest space rookie, do you have any advice for young people who hope to fly on future space missions?" Deke said that the best advice he could give them was "decide what you want to do and then never give up until you've done it!"

Flight Director Neil Hutchinson sagely remarked about working with the Soviets' Flight Director, Vadim Kravets on this flight: "We've gone to all this trouble to learn how to work with those people. It's like going to the Moon once and never going back. 90 percent of the battle is over with...getting all the firsts done...I could run another Apollo-Soyuz or another joint anything with a heck of a lot less fuss than it took to get this one going."

Kubasov remarked later: "Dear American TV people. It would be wrong to ask which country's more beautiful. It would be right to say there is nothing more beautiful than our blue planet."

The Detroit News

U.S.-Soviet handshake in space

Policeman dies; was key witness in narcotics trial

I go to prove my soul!
I see my way as birds their trackless
way.
I shall arrive! what time, what circuit
first,
I ask not; but unless God send his hail
Or blinding fireballs, sleet or stifling snow,
In some time, his good time, I shall
arrive:
He guides me and the bird. In his good
time!

Robert Browning
1812-1889

Mission Control

This emblem was developed to recognize the unique contribution of the Mission Control Team to the Manned Space Program. In selecting the theme of the emblem the Team chose the "Sigma" symbol as the dominant element. Used by Wally Schirra on his Mercury flight, the Sigma represents the total Mission Team, the individual Flight Control Teams from all of the Flights past, present and in the future, all engineering, scientific and operations disciplines and tasks in support of the spacecraft programs. It can also represent many other things such as the benefits for all mankind that are possible through space.

The rocket launch represents the dynamic elements of space, the initial escape from our environment and the thrust to explore the universe.

The remaining elements are the Earth, planets and stars. The Earth is our home and will forever be serviced by both manned and unmanned spacecraft in order to improve the quality of life of our present home. The stars and planets represent a major source of scientific study as well as the challenge of exploration for the future Mission Control Teams. At no time should we lower our sights below the quest for the stars, NASA's explanation goes on, for only in this way are we challenged sufficiently to be better than we are.

The border of the patch contains symbols to represent the three major programs that have been supported by the Team—Mercury, Gemini and Apollo.

The four stars represent the current and future programs: Skylab, Apollo-Soyuz Test Project, Earth Resources Aircraft Program and the Shuttle.

The Latin wording on the patch stands for "Achievement through Excellence."

This patch was designed by famous space artist Robert McCall.

NASA

A gathering of Eagles in the Firing Room during Apollo 10. Left to right: George Low, Manager, Apollo Spacecraft Program; John Williams (hand on head) Director, KSC Spacecraft Operations; Walter Kapryan, Deputy Director, Launch Operations at KSC; General Sam Phillips, Apollo Program Director; Donald "Deke" Slayton, Director of Flight Crew Operations at MSC; and Dr. Kurt Debus, Director, Kennedy Space Center.

Apollo Program officials and Flight Controllers celebrate the successful completion of the Apollo 11 flight in Mission Control with the traditional American flags and post-flight cigars. The Apollo 1 patch can be seen on the far wall above their heads. All of the Astronaut Mission Patches are on the wall in Mission Control.

Mission Control

Dr. Harold C. Urey of the University of California at San Diego had suggested the best landing sites on the Moon for maximum scientific impact. And to those who called the Apollo 11 lunar samples "just a bag of rocks" Urey replied:

> "What a magnificent bag! Rocks last melted 3.65 billion years ago! Dust last chemically assembled 4.66 billion years back at the very beginning of the solar system and of our Mother Earth. We (also) have those marvelous pictures of old Mother Earth as she floats in space."

- ***The men and women of Mission Control made this "bag of rocks" happen.***

Anthropologist Margaret Mead said on the eve of Apollo 11 that it could be "a first step, not into space alone, but into the disciplined and courageous use of enhanced human powers for man, ennobled as he is today, as the first men step on the moon."

- ***The men and women of Mission Control, both "in the trenches and in the backroom" made this "first step" possible.***

And after the Flight of Apollo 11 historian Arthur Schlesinger said: "The 20th Century will be remembered when all else is forgotten, as the century when man burst his terrestrial bonds."

- ***And much of the credit for the bursting of these terrestrial bonds must go to the Men and Women of Mission Control.***

But just as Columbus could not possibly have foreseen the incredible results that would result from his first tentative footsteps in the New World, so, too, Mankind cannot really appreciate or assess the full impact of our first tentative steps into another New World of Outer Space. Dr. Wernher von Braun, who came to America after World War II with 100 of his fellow German engineers and developed America's first space satellite, Explorer, and later the mighty Saturn 5 rocket that took the Astronauts to the Moon, said just before the Flight of Apollo 11:

> "What we will have attained when Neil Armstrong steps down upon the Moon is a completely new step in the evolution of man. For the first time, life will leave its planetary cradle, and the ultimate destiny of Man will no longer be confined to these familiar continents that we have known so long...." Von Braun likened the landing "with aquatic life crawling on Earth for the first time."

- ***Mission Control made sure this would happen.***

Apollo 17 moonwalkers Gene Cernan (center) and Jack Schmitt (right) present the flag that went to the Moon with them on the 1st Anniversary of their flight to Eugene F. Kranz (left) who accepted the flag on behalf of all the Flight Controllers. The ceremony took place during the 3rd Skylab Mission. The Moon Mission flags were made by Annin and Company of Verona, New Jersey.

See America's Space Heritage at these Air & Space Museums

MERCURY CAPSULES

Freedom 7
(Shepard) Air & Space Museum, Washington, D.C.

Liberty Bell 7
(Grissom) Recovered from the Atlantic Ocean floor and now on display at the Kansas Cosmosphere & Space Center, Hutchinson, Kansas.

Friendship 7
(Glenn) Air & Space Museum, Washington, D.C.

Aurora 7
(Carpenter) Museum of Science and Industry, Chicago, IL

Sigma 7
(Schirra) Alabama Space & Rocket Center, Huntsville, Alabama

Faith 7
(Cooper) Johnson Space Center, NASA, Houston, Texas

GEMINI SPACECRAFT

Gemini 3 (Molly Brown)
(Grissom, Young) Grissom Memorial, Spring Mill State Park, Mitchell, Indiana. This was Gus' home town, and he became interested in engineering staring at the water wheel of the old grist mill still in operation there.

Gemini IV
(NASA changed to Roman numerals)
(McDivitt, White) Air & Space Museum, Washington, D.C.

Gemini V
(Cooper, Conrad) Johnson Space Center, NASA, Houston, Texas

Gemini VII
(Borman, Lovell) Air & Space Museum, Washington, D.C.

Gemini VIII
(Armstrong, Scott) Armstrong Museum, Wapakoneta, Ohio

Gemini IX
(Stafford, Cernan) Kennedy Space Center, Florida

Gemini X
(Young, Collins) Kansas Cosmosphere & Space Center, Hutchinson, Kansas

Gemini XI
(Conrad, Gordon) California Museum of Science & Industry, Los Angeles, CA

Gemini XII
(Lovell, Aldrin) NASA Goddard Space Flight Center, Greenbelt, MD

Gemini VI
(Schirra, Stafford) St. Louis Science Center, St. Louis, Mo.

Time does not diminish the awe felt at the audacity of Man's first small steps into the cosmos. National Air & Space Museum, Washington, D.C. This was America's first spaceship, piloted by Alan Shepard. May 5, 1961. Photo by the author.

APOLLO SPACECRAFT

Apollo 1
(Grissom, White, Chaffee) Fatal Fire, Not on Display

Apollo 7
(Schirra, Cunningham, Eisele) National Museum of Science & Technology, Ottawa, Canada

Apollo 8
(Borman, Lovell, Anders) Chicago Museum of Science & Industry, Chicago, Illinois

Apollo 9
(McDivitt, Scott, Schweickart) Michigan Space Center, Jackson Community College, Jackson, Michigan

Apollo 10
(Stafford, Young, Cernan) Science Museum, London, England

Apollo 11
(Armstrong, Aldrin, Collins) Air & Space Museum, Washington, D.C.

Apollo 12
(Conrad, Bean, Gordon) Langley Research Center, NASA, Langley, Virginia

Apollo 13
(Lovell, Swigert, Haise) Kansas Cosmosphere & Space Center, Hutchinson, Kansas

Apollo 14
(Shepard, Mitchell, Roosa) Rockwell International, Downey, California

Apollo 15
(Scott, Worden, Irwin) Air Force Museum, Wright-Patterson Air Force Base, Dayton, Ohio

Apollo 16
(Young, Mattingly, Duke) Alabama Space & Rocket Center, Huntsville, Alabama

Apollo 17
(Cernan, Evans, Schmitt) Johnson Space Center, NASA, Houston, Texas

A mighty Saturn V aft engine dwarfs a young observer. There were five of these engine bells on the end of the rocket that took us to the Moon. National Air & Space Museum, Washington, D.C. Photo by the author.

SKYLAB COMMAND MODULES

Skylab 1

(Conrad, Weitz, Kerwin) Naval Aviation Museum, Pensacola, Florida

Skylab 2

(Bean, Lousma, Garriott) NASA Lewis Research Center, Cleveland, OH

Skylab 3

(Carr, Pogue, Gibson) National Air & Space Museum, Washington, D.C.

APOLLO-SOYUZ TEST PROJECT

Apollo Command Module

(Stafford, Brand, Slayton) Kennedy Space Center, Florida

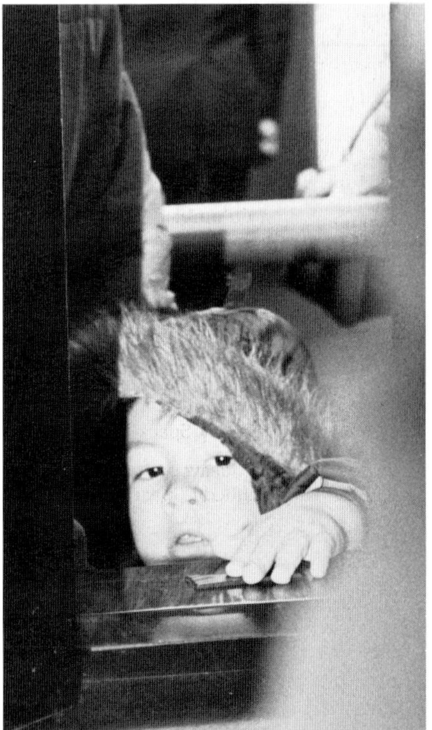

Doing the impossible. A youngster touches an actual piece of the Moon. National Air & Space Museum, Washington, D.C. Photo by the author.

SPACE TRANSPORTATION SYSTEM
Space Shuttle Orbiters

The Astronauts who flight-tested the Space Shuttle "Enterprise" in Earth's atmosphere, left to right, Gordon Fullerton, Fred Haise, Joe Engle and Dick Truly.

Enterprise
(Engle, Truly, Haise, Fullerton) (Did Not Fly In Space, Used for Atmospheric Testing)

Columbia
On Active Duty, Kennedy Space Center, Florida

Challenger
Lost During Launch

Discovery
On Active Duty, Kennedy Space Center, Florida

Atlantis
On Active Duty, Kennedy Space Center, Florida

Endeavour
On Active Duty, Kennedy Space Center, Florida

Sources: Science Digest, September 1981, © The Hearst Corporation; Committee on Science & Technology Report, Library of Congress, Congressional Research Service, 1981; America's Journeys into Space, Anthony J. Cipriano, Wanderer Books, New York.

Note: There are more than 50 Aviation and Space Museums in the United States and Canada. A complete list is available in "Aviation and Space Museums of America" by Jon L. Allen, Arco Publishing Company, Inc., 219 Park Avenue South, New York, N.Y. 10003.

Kennedy Space Center Tours

The waiting Shuttle at night before launch.

The first rocket was launched from Cape Canaveral on July 24, 1950. The author visited the area barely one month later and was hooked forever on the idea of space travel and rocketry. A cement slab served as the launch pad for that first launch and the control center was a tarpaper shack that had once been a bath house. The vehicle was a 56 foot WAC Bumper, a modified German V-2 brought to this country by Wernher von Braun, Dr. Kurt Debus and their friends. Colonel Harold R. Turner headed up the firing team. The rocket climbed 10 miles above Earth.

Today you will probably see a Space Shuttle sitting on its launch pad when you visit the Kennedy Space Center. Now that America has a fleet of them on-stream chances are good that everyone that wants to will be able to see a launch.

Bus tours of the Kennedy Space Center are conducted every day of the year except Christmas. Tour routes and departure schedules may vary due to launch or landing operations. An entire day can be spent attending theatre presentations, browsing through space-related exhibits or taking the bus tours. Not to be missed is the spectacular movie "The Dream Is Alive." Exhibits include actual rockets, spacecraft and scale models of a variety of NASA space vehicles.

Meet a real astronaut at the popular Astronaut Encounter Show at the Kennedy Space Center Visitor Complex.

- **KSC Tour.** This tour takes visitors to restricted areas of NASA including the International Space Station Center where they see real components of the Space Station being readied for launch; the Apollo/Saturn V Center where they walk under an actual 363-foot Saturn V rocket; and the LC 39 Observation where they will be within miles of Launch Pads 39A and B.

- **Mad Mission to Mars 2025.** This new, live stage show uses wacky characters to make learning about America's space program loads of fun. At the end of the show, visitors take a virtual trip to Mars!

- **Astronaut Encounter Show.** This live show allows guests to talk with a real astronaut every day of the year! Kennedy Space Center Visitor Complex is the only place in the world where guests can interact with one of the few who has flown in space.

For more information, write:

Kennedy Space Center Visitor Complex
Mail Code: DNPS
Kennedy Space Center, FL 32899

Tourbooks are available in English, German, Portuguese, Spanish and French.

Free kennel service while you are on tour.

For launch info, call 1-800-432-2153
For tour info, call (321) 449-4444

Johnson Space Center Tours

A portion of the epic mural by noted space artist Robert McCall in the lobby of the Visitor Information Center at the Johnson Space Center.

The Johnson Space Center is located 25 miles southeast of Houston, Texas on former pasture land then populated by wolves, wild turkeys, deer and fox. Its 1620 acres opened for the space business in September 1963. This is where the astronauts live and train. Here is Mission Control, the Shuttle Training Simulators, the Lunar Receiving Laboratory, and much much more.

Space Center Houston, the official visitor center for the Johnson Space Center, offers a variety of exhibits and tours for the public. Located outside the main gates of the Johnson Space Center, they offer a guided tram tour of Mission Control and astronaut training facilities. For additional information on tours, exhibits and operating hours, visit their website at www.spacecenter.org or call Space Center Houston at 281-244-2105.

Hours of Operation:	Labor Day to Memorial Day: Monday–Friday: 10:00 a.m. to 5:00 p.m. Saturday–Sunday: 10:00 a.m. to 7:00 p.m.
	Memorial Day to Labor Day: Daily 9:00 a.m. to 7:00 p.m.
	Open Daily Except Christmas
Admission Charge:	Adults: $14.95 Seniors: $13.95 Children: $10.95
Group Sales:	281-244-2130 (15 or more)
	Tours requiring special arrangements can call 281-244-2115

U.S. SPACE CAMP®
Youth Science Program

Consulting flight manuals and flipping switches are routine tasks for trainees who are assigned to commander and pilot positions inside a shuttle cockpit during a simulated mission. (Photo by Charles Seifried/U.S. Space & Rocket Center)

Conducted by the U.S. Space & Rocket Center, Huntsville, Alabama *(Open daily except Christmas).* Camps are also located in Titusville, Florida and Mountain View, California and four other countries. Aviation Challenge, a similar program that allows students to train like fighter pilots, is available in Huntsville and Atwater, California.

Age: Open to students in grades 4–12. Programs available for teachers and other adults.

Time: Sessions are scheduled year-round. Five-day and three-day programs available.

Cost: Varies, depending on age level and length of program.

Accommodations: Space Camp Habitat—a three-level dormitory that was designed similar to a space station.

It's a week of astronaut training, teamwork, decision-making, problem-solving and learning how math and science are applied in the space program. Students experience simulators and conduct simulated space shuttle missions. Featured in the movie "Space Camp" and acclaimed worldwide by The New York Times, "Good Morning, America," The National Geographic Society, "Today," The Learning Channel, "The Oprah Winfrey Show," The Weekly Reader and media around the world.

For more information: Call toll-free 1-800-63-SPACE or visit the web site at www.spacecamp.com.

The first view of Mars, taken by Viking 1 on July 20, 1976. It is late in the Martian afternoon. The horizon is 1.8 miles away. The surface of Mars awaits Mankind.

THE INDIANAPOLIS STAR

TUESDAY, SEPTEMBER 16, 1969

U.S. ON MARS IN '80s: NIXON

Thant Asks Russ-China-U.S. Accord

5 'A-Powers' Urged To Join In Disarm Talks

G.I.s Gone, Reds Push Into Delta

Kickback Probe Jury, Hears Hunt Testimony

President OKs Space Goal Set By Agnew Group

Ulster Gives Warning On Barricades

AFTERWORD

Howard Benedict: (Associated Press) — Are you suggesting Space Exploration as a substitute for war?

Dr. George E. Mueller: (NASA Associate Administrator) — If it is pursued as actively as we pursue wars between nations, we will be in a position to solve that basic drive which does lead to wars, to sublimate it with a drive to conquer space, conquer other planets and eventually travel from this solar system to other star systems. I think the next generation not only is going to accept, but expect the continuation of Space Exploration as part of this nation and part of the world. I can think of no better way of draining off the creative energies of the world."

Editor's Note: The United States has not moved forward in attempting to reach President Nixon's Goal as announced in this September 16, 1969 headline. Perhaps the lessons of Apollo-Soyuz and Dr. Mueller's comments above have merit. Perhaps we need a joint United Nations Manned Expeditionary Mission to Mars composed of an American, a Russian and a Chinese—ARC. A Covenant of the ARC.

Compiled by Michael Lattimer.

APOLLO CONTRACTORS

We salute the men and women here in America and elsewhere around the world who played such an important role in making our successful moon landings "happen." In the following list the primary contractor is shown in capital letters, followed by those companies and agencies that also played a major role in each component in the Apollo Program.

Command & Service Module and Lunar Adapter

NORTH AMERICAN
ROCKWELL CORP.
DOWNEY, CALIFORNIA

Aerojet-General Corp.
Sacramento, California

Aeronca, Inc.
Middletown, Ohio

Air Products & Chemicals
Allentown, Pennsylvania

Applied Electronics Corp.
Metuchen, New Jersey

Avco Corporation
Lowell, Massachusetts

Avien, Inc.
Woodside, New York

Beckman Instruments, Inc.
Fullerton, California

Beech Aircraft Corp.
Boulder, Colorado

Bell Aerospace Corp.
Buffalo, New York

Bendix Corporation
Teterboro, New Jersey

Carleton Controls Corp.
East Aurora, New York

Clevite Corporation
Cleveland, Ohio

Collins Radio Company
Cedar Rapids, Iowa

Conrac Corporation
Pasadena, California

Control Data Corporation
Minneapolis, Minnesota

Cosmodyne Corporation
Torrance, California

ESB, Inc.
Raleigh, North Carolina

Eagle-Picher Industries
Joplin, Missouri

Ex-cell-o Corporation
Costa Mesa, California

Fairchild Hiller Corp.
Manhattan Beach, California

The Garrett Corporation
Los Angeles, California

General Motors Corporation
Milwaukee, Wisconsin and
Indianapolis, Indiana

General Time Corporation
Rolling Meadows, Illinois

Gibbs Mfg. & Research Corp.
Janesville, Wisconsin

Honeywell, Inc.
Minneapolis, Minnesota

ILC Industries, Inc.
Dover, Delaware

International Telephone &
Telegraph Corporation
San Fernando, California and
Chicago, Illinois

Weber Aircraft, Kidde, Inc.
Burbank, California

Kollsman Instrument Corp.
Syossett, New York

Leach Corporation
Azusa, California

Lear Siegler, Inc.
Elyria, Ohio

Ling-Temco-Vought, Inc.
Dallas, Texas

Litton Systems, Inc.
College Park, Maryland

The Marquardt Corporation
Van Nuys, California

Massachusetts Institute
of Technology
Cambridge, Massachusetts

McGraw-Edison Company
Livingston, New Jersey

Microdot, Inc.
South Pasadena, California

Motorola, Inc.
Scottsdale, Arizona

North American Rockwell Corp.
Anaheim, California
Los Angeles, California
Canoga Park, California
Tulsa, Oklahoma

Northrop Corporation
Newbury Park, California

Opcalite, Inc.
Santa Ana, California

Parker-Hannifin Corporation
Los Angeles, California

Philco-Ford Corporation
Palo Alto, California

Pneumo Dynamics Corporation
Kalamazoo, Michigan

RCA Corporation
Princeton, New Jersey

Radiation, Inc.
Melbourne, Florida

Raytheon Company
Sudbury, Massachusetts

Rosemount Engineering Co.
Minneapolis, Minnesota

Royal Industries
Pomona, California

Sargent Industries, Inc.
Los Angeles, California

Sciaky Brothers, Inc.
Chicago, Illinois

Simmonds Precision Products
Tarrytown, New York

Space Ordnance Systems
El Segundo, California

Sperry Rand Corporation
Great Neck, New York

Teledyne, Inc.
Solana Beach, California

Textron, Inc.
Belmont, California
Whittier, California

Transco Products, Inc.
Venice, California

United Aircraft Corporation
Windsor Locks, Connecticut
Hartford, Connecticut

Westinghouse Electric Corp.
Baltimore, Maryland and
Lima, Ohio

Weston Instruments, Inc.
Newark, New Jersey

Whittaker Corporation
North Hollywood, California

Xerox Corporation
Pasadena, California

Mission Control and Evaluation

Bendix Corporation
Ann Arbor, Michigan
Southfield, Michigan

Brown & Root/Northrop
Houston, Texas

Cohu Electronics, Inc.
San Diego, California

General Dynamics Corporation
Houston, Texas
Orlando, Florida

International Business Machines
Houston, Texas

Ling-Temco-Vought, Inc.
Dallas, Texas

Little, A.D. Co.
Cambridge, Massachusetts

Lockheed Aircraft Corporation
Plainfield, New Jersey

Melpar, Inc.
Falls Church, Virginia

Philco-Ford Corporation
Palo Alta, California

Sperry Rand Corporation
Philadelphia, Pennsylvania

TRW, Inc.
Redondo Beach, California

Teledyne, Inc.
Garland, Texas
Pasadena, California

Launch Escape System

NORTH AMERICAN
ROCKWELL CORP.
DOWNEY, CALIFORNIA

Lockheed Aircraft Corporation
Redlands, California

North American Rockwell Corp.
Inglewood, California

Thiokol Chemical Corporation
Elkton, Maryland

Lunar Module

GRUMMAN AEROSPACE
CORPORATION
BETHPAGE, NEW YORK

AMBAC Industries, Inc.
Garden City, New York

Aerojet-General Corporation
Downey, California

Avco Corporation
Cincinnati, Ohio

Bell Aerospace Corporation
Buffalo, New York

Collins Radio Company
Dallas, Texas

Conrac Corporation
Duarte, California

Corning Glass Works
Corning, New York

Dynamics Corp. of America
Garden City, New York

Eagle-Picher Industries, Inc.
Joplin, Missouri

Eastman Kodak Company
Rochester, New York

Edo Corporation
College Point, New York

Farrand Optical Co., Inc.
New York, New York

The Garrett Corporation
Los Angeles, California

General Electric Company
Waynesboro, Virginia

General Motors Corporation
Indianapolis, Indiana and
Milwaukee, Wisconsin

HITCO
Los Angeles, California

Honeywell, Inc.
Minneapolis, Minnesotas

ILC Industries, Inc.
Dover, Delaware

Kollsman Instrument Corp.
Syossett, New York

Leach Corporation
Azusa, California

Lear Siegler, Inc.
Grand Rapids, Michigan

The Marquardt Corporation
Van Nuys, California

Massachusetts Institute
of Technology
Cambridge, Massachusetts

Motorola, Inc.
Scottsdale, Arizona

North American
Rockwell Corporation
Canoga Park, California

RCA Corporation
Burlington, Massachusetts
Camden, New Jersey

Radiation, Inc.
Palm Bay, Florida

Raytheon Company
Lexington, Massachusetts

Ryan Aeronautical Company
San Diego, California

Sargent Industries, Inc.
El Segundo, California

Sylvania Electric Products, Inc.
Needham Heights, Massachusetts

TRW, Inc.
Redondo Beach, California

Textron, Inc.
Belmont, California

Trans-Sonics, Inc.
Lexington, Massachusetts

United Aircraft Corporation
Windsor Locks, Connecticut
East Hartford, Connecticut

Westinghouse Electric Corp.
Pittsburgh, Pennsylvania

Instrument Unit

INTERNATIONAL
BUSINESS MACHINES
GAITHERSBURG, MARYLAND

Aerodyne Controls Corporation
Farmingdale, New York

Avco Corporation
Huntsville, Alabama
Nashville, Tennessee

Bendix Corporation
Teterboro, New Jersey

Eagle-Picher Industries, Inc.
Joplin, Missouri

Electro Development Corporation
Seattle, Washington

Electronic Communications, Inc.
St. Petersburg, Florida

Fansteel, Inc.
Compton, California

Martin Marietta Corporation
Orlando, Florida

Motorola, Inc.
Scottsdale, Arizona

North American
Rockwell Corporation
Tulsa, Oklahoma

Northrop Corporation
Norwood, Massachusetts

Brown Engineering Co., Inc.
Huntsville, Alabama

Chrysler Corporation
Huntsville, Alabama

Hayes International Corporation
Huntsville, Alabama

International Harvester Company
San Diego, California

International Telephone and
Telegraph Corporation
Clinton, Massachusetts

Space Craft, Inc.
Huntsville, Alabama

TRW, Inc.
Cleveland, Ohio

United Aircraft Corporation
Windsor Locks, Connecticut

Saturn Rocket–First Stage

THE BOEING COMPANY
SEATTLE, WASHINGTON

Aircraft Porous Media, Inc.
Glen Cove, New York

Allen-Bradley Company
Milwaukee, Wisconsin

Babcock Electronics
Costa Mesa, California

Bell Aerospace Corporation
Burbank, California

Bendix Corporation
Davenport, Iowa

Bourns, Inc.
Riverside, California

Brown Engineering Co., Inc.
Lewisburg, Tennessee

Calumet & Hecla, Inc.
Bartlett, Illinois

The J.C. Carter Company
Costa Mesa, California

Consolidated Controls Corp.
Bethel, Connecticut and
Los Angeles, California

The Deutsch Company
Long Island, New York

Dickson Electronics Corporation
Scottsdale, Arizona

Dynasciences Corporation
Chatsworth, California

Eagle-Picher Industries, Inc.
Joplin, Missouri

Electro Development Corporation
Seattle, Washington

Federal-Mogul Corporation
Los Alamitos, California

Flexible Tubing Corporation
Anaheim, California

The Garrett Corporation
Phoenix, Arizona

General Electric Company
Auburn, New York

Hayes International Corporation
Birmingham, Alabama

International Harvester Company
San Diego, California

International Telephone and
Telegraph Corporation
Los Angeles, California

Johns-Manville Sales Corporation
Manville, New Jersey

LTV Aerospace Corporation
Dallas, Texas

Marotta Scientific Controls
Boonton, New Jersey

Martin Marietta Corporation
Baltimore, Maryland

Mepco, Inc.
Morristown, New Jersey

Methode Electronics, Inc.
Chicago, Illinois

Moog, Inc.
East Aurora, New York

Navan, Inc.
El Segundo, California

Parker-Hannifin Corporation
Los Angeles, California

Parker Seal Company
Culver City, California

Parsons Corporation
Traverse City, Michigan

Purolator Products, Inc.
Newbury Park, California

Raytheon Company
Mountain View, California
Wayland, Massachusetts

Resistoflex Corporation
Roseland, New Jersey

Rohr Corporation
Chula Vista, California

Sargent Industries, Inc.
El Segundo, California

Servonic Instruments, Inc.
Costa Mesa, California

Sprague Electric Company
North Adams, Massachusetts

Stainless Steel Products
Burbank, California

Sterer Engineering and
Manufacturing Company
Los Angeles, California

Teledyne, Inc.
Solana Beach, California

Thiokol Chemical Corporation
Elkton, Maryland

Trans-Sonics, Inc.
Burlington, Massachusetts

UMC Industries, Inc.
St. Louis, Missouri

Westinghouse Air Brake Company
Pittsburgh, Pennsylvania

Wright, Fred D. Company
Nashville, Tennessee

Saturn Rocket–Second Stage

NORTH AMERICAN
ROCKWELL CORPORATION
DOWNEY, CALIFORNIA

Abex Corporation
Oxnard, California

Acoustics Associates, Inc.
Los Angeles, California

Ametek, Inc.
El Cajon, California
Los Angeles, California

Anaconda American Brass Co.
Los Angeles, California

The Deutsch Company
Banning, California

Eagle-Picher Industries, Inc.
Joplin, Missouri

Explosive Technology
Fairfield, California

Futurecraft Corporation
City of Industry, California

Goodyear Aerospace Corporation
Litchfield Park, Arizona

Gulton Industries, Inc.
Hawthorne, California

Parker-Hannifin Corporation
Los Angeles, California

RdF Corporation
Hudson, New Hampshire

Resistoflex Corporation
Roseland, New Jersey

Servonic Instruments, Inc.
Costa Mesa, California

Snap-Tite, Inc.
Union City, Pennsylvania

Space Craft, Inc.
Huntsville, Alabama

Babcock Electronics
Costa Mesa, California

Barry Wright Corporation
Glendale, California

Borg-Warner Corporation
Bedford, Ohio

Brown Engineering Co., Inc.
Lewisburg, Tennessee

International Harvester Company
San Diego, California

Lord Corporation
Erie, Pennsylvania

North American
Rockwell Corporation
McGregor, Texas

Stainless Steel Products
Burbank, California

Teledyne, Inc.
Solana Beach, California

Transco Products, Inc.
Venice, California

UMC Industries, Inc.
Goodyear, Arizona

Saturn Rocket–Third Stage

McDONNEL DOUGLAS CORP.
HUNTINGTON BEACH,
CALIFORNIA

Adel/Vinson
Burbank, California

Aeroquip Corporation
Los Angeles, California

Airdrome Parts Company
Inglewood, California

American Machine & Foundry
Princeton, Indiana

Ametek, Inc.
Los Angeles, California

Anaconda American Brass Co.
Los Angeles, California

Avnet Electronics Corporation
Culver City, California

Barry Wright Corporation
Watertown, Massachusetts

Bell Aerospace Corporation
Buffalo, New York

Bell & Howell Company
New York, New York

Bertea Products
Irvine, California

Borg-Warner Corporation
Bedford, Ohio

Bourns, Inc.
Riverside, California

Brown Engineering Co., Inc.
Huntsville, Alabama

Bunker-Ramo Corporation
Chicago, Illinois

Carlton Industries
Hawthorne, California

Data Sensors, Inc.
Gardena, California

The Deutsch Company
Los Angeles, California

Eagle-Picher Industries, Inc.
Joplin, Missouri

Fairchild Camera & Instrument
Montebello, California
Hollywood, California

Fairchild Hiller Corporation
Bayshore, New York
Manhattan Beach, California

Fansteel, Inc.
Compton, California

Flexible Metal Hose Mfg. Co.
Northridge, California

Frebank Company
Glendale, California

The Garrett Corporation
Phoenix, Arizona

Grove Valve & Regulator Co.
Oakland, California

Gulton Industries, Inc.
Fullerton, California

Honeywell, Inc.
Minneapolis, Minnesota

Leonard, Wallace O., Inc.
Pasadena, California

Litton Industries, Inc.
Beverly Hills, California

Magnesium Alloy Prods. Co.
Compton, California

Magnetika, Inc.
Venice, California

Marotta Scientific Controls
Santa Ana, California

Menasco Mfg. Co.
Burbank, California

Moog, Inc.
East Aurora, New York

Motorola, Inc.
Hollywood, California

Parker-Hannifin Corp.
Los Angeles, California

Purolator Products, Inc.
Van Nuys, California

Rosemount Engineering Co.
Minneapolis, Minnesota

Royal Industries
Pomona, California

Sealol, Inc.
Providence, Rhode Island

Signet Scientific Co.
Burbank, California

Sperry Rand Corporation
Troy, Michigan

Stainless Steel Products
Burbank, California

Statham Instruments, Inc.
Los Angeles, California

TRW, Inc.
Cleveland, Ohio

Technology Instrument Corp.
Newbury Park, California

Teledyne, Inc.
Costa Mesa, California
Solana Beach, California

Textron, Inc.
Whittier, California

Texas Instruments, Inc.
Dallas, Texas

Thiokol Chemical Corporation
Elkton, Maryland

Trans-Sonics, Inc.
Lexington, Massachusetts

Vacco Industries
South El Monte, California

Vinson Mfg. Co., Inc.
Van Nuys, California

Saturn Rocket–First Stage Engine

NORTH AMERICAN
ROCKWELL CORPORATION
CANOGA PARK, CALIFORNIA

A&M Castings, Inc.
South Gate, California

Ace Industries
Santa Fe Springs, California

Clevite Corporation
Cleveland, Ohio

Fairchild Camera & Instrument
El Cajon, California

L.A. Gauge Co., Inc.
Sun Valley, California

Parker Seal Company
Culver City, California

Paragon Tool, Die & Eng.
Pacoima, California

Precision Sheet Metal, Inc.
Los Angeles, California

Adept Manufacturing Company
Los Angeles, California

Anaconda American Brass Company
Detroit, Michigan
Los Angeles, California

Arcturus Manufacturing Company
Oxnard, California

Bell & Howell Company
Pasadena, California

Bendix Corporation
Sidney, New York

Beuhler Corporation
Indianapolis, Indiana

Borg-Warner Corporation
Vernon, California

Bunker-Ramo Corporation
Broadview, Illinois

Calif.-Doran Heat Treating
Los Angeles, California

Cam Car Company
Rockford, Illinois

Chicago Rawhide Mfg. Co.
Chicago, Illinois

General Laboratory Associates
Norwich, New York

Howmet Corporation
Dover, New Jersey

Industrial Tectonics, Inc.
Ann Arbor, Michigan

International Harvester Company
San Diego, California

International Nickel Co., Inc.
Huntington, West Virginia

International Telephone and
Telegraph Corporation
Boston, Massachusetts

Keystone Engineering Co.
Los Angeles, California

Langley Corporation
San Diego, California

LeFeill Manufacturing Co.
Santa Fe Springs, California

MSL Industries, Inc.
Batavia, Illinois
North Hollywood, California

McWilliams Forge Company
Rockaway, New Jersey

National Banner
Los Angeles, California

Quadrant Eng. Company
Gardena, California

Reisner Metals, Inc.
South Gate, California

Rohr Corporation
Chula Vista, California

Southwestern Industries, Inc.
Los Angeles, California

Standard Pressed Steel Company
Santa Ana, California
Jenkinstown, Pennsylvania

Statham Instruments, Inc.
Los Angeles, California

Texas Instruments, Inc.
Dallas, Texas

Union Carbide Corporation
Kokomo, Indiana

Viking Forge & Steel Company
Albany, California

Western Way, Inc.
Van Nuys, California

Wyman-Gordon Company
N. Grafton, Massachusetts

Saturn Rocket–Second & Third Stage Engines

NORTH AMERICAN
ROCKWELL CORPORATION
CANOGA PARK, CALIFORNIA

A&M Castings, Inc.
South Gate, California

Anaconda American Brass Co.
Detroit, Michigan
Los Angeles, California

Arcturus Manufacturing Co.
Oxnard, California

Bell & Howell Company
Pasadena, California

Bendix Corporation
Sidney, New York

Bunker-Ramo Corporation
Broadview, Illinois

Calif.-Doran Heat Treating
Los Angeles, California

Cam Car Company
Rockford, Illinois

Chicago Rawhide Mfg. Co.
Chicago, Illinois

Clevite Corporation
Cleveland, Ohio

Fairchild Camera & Instrument
El Cajon, California

L.A. Gauge Co., Inc.
Sun Valley, California

General Laboratory Associates
Norwich, New York

Howmet Corporation
Dover, New Jersey

Industrial Tectonics, Inc.
Ann Arbor, Michigan

International Harvester Company
San Diego, California

International Nickel Company
Huntington, West Virginia

International Telephone and
Telegraph Corporation
Boston, Massachusetts

Kentucky Metals, Inc.
Louisville, Kentucky

Keystone Engineering Co.
Los Angeles, California

Langley Corporation
San Diego, California

LeFeill Mfg. Company
Santa Fe Springs, California

MSL Industries, Inc.
North Hollywood, California
Batavia, Illinois

McWilliams Forge Company
Rockaway, New Jersey

National Banner
Los Angeles, California

Parker Seal Company
Culver City, California

Quadrant Engrg. Company
Gardena, California

Reisner Metals, Inc.
South Gate, California

Scientific Data Systems
Pomona, California

Southwestern Industries, Inc.
Los Angeles, California

Standard Pressed Steel Co.
Santa Ana, California
Jenkinstown, Pennsylvania

Statham Instruments, Inc.
Los Angeles, California

Texas Instruments, Inc.
Dallas, Texas

Union Carbide Corporation
Kokomo, Indiana

Viking Forge & Steel Co.
Albany, California

Western Way, Inc.
Van Nuys, California

Wyman-Gordon Company
N. Grafton, Massachusetts

AUTHOR'S NOTE: The J-2 Engine was used on both the second and third stages of the Saturn stack: a cluster of five on the second stage and a single gimbal-mounted J-2 Engine on the third stage so that it could be moved in flight to steer the stage toward the Moon.

Communications and Tracking and Data Acquisition

Amalgamated Wireless Australasia
Carnarvon, Australia

Australia, Commonwealth of
Department of Works
Canberra, Australia

Australia, Commonwealth of
Postmaster General
Canberra, Australia

American Telephone & Telegraph
Washington, D.C.

Ampex Corporation
Redwood City, California

Avondale Shipyard
Algiers, Louisiana

Bell & Howell Company
Arlington, Virginia

Bendix Field Engineering Corp.
Owings Mills, Maryland

Blaw-Knox Company
Pittsburgh, Pennsylvania

Cable & Wireless Limited
London, England

Canoga Electronics Corporation
Chatsworth, California

Chesapeake & Potomac
Telephone Co.
Washington, D.C.

Clevite Corporation
Cleveland, Ohio

Collins Radio Company
Dallas, Texas

Commonwealth and Industrial
Scientific Research Organization
Parkes, Australia

Communications Satellite Corp.
Washington, D.C.

Compania Telefonica Nacional
De Espana
Madrid, Spain

Computing and Software, Inc.
Panorama City, California

DBA Systems, Inc.
Melbourne, Florida

Dynamics Corp. of America
Garden City, New York

Electrica Newbery del Guaymas
Snonoral, Mexico

Emerson Electric Company
Calabasas, California

Energy Systems, Inc.
Palo Alto, California

Extreme Telecommunications
Executive
London, England

Federal Electric Corporation
Paramus, New Jersey

General Dynamics Corporation
San Diego, California
Orlando, Florida
Quincy, Massachusetts

General Instrument Corporation
Westwood, Massachusetts

Hawaiian Telephone Company
Honolulu, Hawaii

Hewlett-Packard Company
Palo Alto, California

ITT World Communications, Inc.
Washington, D.C.

Institute Nacional de Tecnica
Aeroespacial
Madrid, Spain

International Business Machines
Gaithersburg, Maryland
Poughkeepsie, New York

Jet Propulsion Laboratory
Pasadena, California

Johns Hopkins University
Silver Spring, Maryland

LTV Aerospace Corporation
Honolulu, Hawaii

Ling-Temco-Vought, Inc.
Dallas, Texas

3 M Company
Camarillo, California

Milgo Electronic Corporation
Miami, Florida

Motorola, Inc.
Scottsdale, Arizona

Overseas Telecommunications
Commission
Sydney, Australia

Philco-Ford Corporation
Fort Washington, Pennsylvania

Potter Instrument Co., Inc.
Plainview, New York

Precision Fabricators
Paramount, California

RCA Corporation
Morristown, New Jersey
Princeton, New Jersey
Cherry Hill, New Jersey
New York, New York

Raytheon Company
Waltham, Massachusetts

Space Track Limited
Canberra, Australia

Sperry Rand Corporation
Great Neck, New York
St. Paul, Minnesota

Standard Telephones & Cables, Ltd.
Essex, England

Stellametrics, Inc.
Santa Barbara, California

TRW, Inc.
Redondo Beach, California

Teletype Corporation
Skokie, Illinois

Varian Associates
Palo Alto, California

Vitro Corp. of America
Silver Spring, Maryland

Western Gear Corporation
Everett, Washington

Western Union International
New York, New York

Westinghouse Electric Corporation
Baltimore, Maryland

Weston Instruments, Inc.
College Park, Maryland

Ordnance Vendors

Aerojet-General Corporation
Sacramento, California

Ensign-Bickford Company
Simsbury, Connecticut

Explosive Technology
Fairfield, California

General Laboratory Associates
Norwich, New York

Teledyne, Inc.
Hollister, California

Launch Support

Air Products & Chemicals, Inc.
Allentown, Pennsylvania

Akwa-Downey Construction Co.
Milwaukee, Wisconsin

American Elcon, Inc.
Cape Canaveral, Florida

American Machine & Foundry Co.
Santa Barbara, California

Ametek, Inc.
El Cajon, California

Ampex Corporation
Redwood City, California
Cocoa Beach, Florida

Astrodata, Inc.
Anaheim, California

Automatic Retailers of America
Philadelphia, Pennsylvania

Bechtel Corporation
San Francisco, California

Beckman Instruments, Inc.
Fullerton, California

Bell & Howell Company
Pasadena, California

Bendix Corporation
Cocoa Beach, Florida

Big Three Industrial Gas
& Equipment Company
Houston, Texas

Blount Brothers Corporation and
M.M. Sundt Construction Company
Montgomery, Alabama

Boeing Company
Seattle, Washington

Brevard Engineering Company
Cape Canaveral, Florida

Brown Engineering Co., Inc.
Huntsville, Alabama

Carrier Corporation
Syracuse, New York

Catalytic Construction Company
Philadelphia, Pennsylvania

Catalytic-Dow
Titusville, Florida

Chicago Bridge & Iron Company
Chicago, Illinois

R.E. Clarson, Inc.
St. Petersburg, Florida

Collins Radio Company
Richardson, Texas

Cosmodyne Corporation
Torrance, California

Data-Control Systems, Inc.
Danbury, Connecticut

Arnold M. Diamond, Inc.
Great Neck, New York

Dow Chemical Company
Midland, Michigan

Dyna-Therm Corporation
Los Angeles, California

Ernst/Smith, Joint Venture
Orlando, Florida

Howard O. Foley Company
Washington, D.C.

Gahagan Dredging Corporation
Tampa, Florida

General Dynamics Corporation
San Diego, California

General Electric Company
Huntsville, Alabama

Giffels and Rossette
Detroit, Michigan

Greening and Sayer
Daytona Beach, Florida

Grumman Aerospace Corporation
Bethpage, New York

Gulton Industries, Inc.
Albuquerque, New Mexico

Hayes International Corporation
Birmingham, Alabama

Heyl & Patterson, Inc.
Cocoa, Florida

Honeywell, Inc.
Orlando, Florida

Ingalls Iron Works Company
Birmingham, Alabama

International Business Machines
Gaitherburg, Maryland

International Telephone and
Telegraph Corporation
Cocoa Beach, Florida

Kahn & Company
Hartford, Connecticut

Kaminer Construction Corporation
Chamblee, Georgia

Litton Industries, Inc.
Pittsburgh, Pennsylvania

Marion Power Shovel Company
Marion, Ohio

C.A. Mayer Paving &
Construction Co.
Orlando, Florida

McDonnel Douglas Corporation
Santa Monica, California

McGregor and Warner, Inc.
Washington, D.C.

Molecular Research, Inc.
West Palm Beach, Florida

Monitor System, Inc.
Fort Washington, Pennsylvania

Morrison-Knudsen Co., Perini
Corp., Paul Hardeman, Inc.
Joint Venture
South Gate, California

Natkin & Company
Kansas City, Missouri

North American Rockwell Corp.
Downey, California

Otis Elevator Company
New York, New York

Pacific Crane & Rigging Company
Paramount, California

W.V. Pangborne Co. and Lowry
Electric Company–Joint Venture
Philadelphia, Pennsylvania

Philpott, Ross & Saarinen, Inc.
Fort Lauderdale, Florida

RCA Corporation
Van Nuys, California
Camden, New Jersey

Remler Company
Sam Francisco, California

Reynolds, Smith & Hills
Jacksonville, Florida

Rogers, Lovelock and Fritz
Winter Park, Florida

Rosemount Engineering Company
Minneapolis, Minnesota

Sanders Associates, Inc.
Nashua, New Hampshire

Scientific Data Systems
Orlando, Florida

Service Technology Corporation
Dallas, Texas

Smith & Sapp Construction Co.
Orlando, Florida

Space Corporation
Dallas, Texas

Statham Instruments, Inc.
Los Angeles, California

Symetrics Engineering Corporation
of Florida
Satellite Beach, Florida

Technicolor Corporation
Cocoa Beach, Florida

Tektronix, Inc.
Orlando, Florida

Trans World Airlines, Inc.
Titusville, Florida

Union Carbide Corporation
Tonawanda, New York

United States Steel Corporation
Pittsburgh, Pennsylvania

We Believe
in every young mind there is a
Window on the Universe

Challenger Center is a global not-for-profit education organization created in 1986 by the families of the astronauts tragically lost during the last flight of the Challenger space shuttle.

Dedicated to the educational spirit of that mission, Challenger Center develops Learning Centers and other educational programs worldwide to continue the mission to engage students in science and math education.

We invite you to join our mission.

Challenger Center for Space Science Education
1250 North Pitt Street
Alexandria, VA 22314

(703) 683-9740 • *www.challenger.org*